Soil Characterization, Classification and Survey

Soil Characterization, Classification and Survey

Ivara Ejemot Esu

 HEBN Publishers Plc

HEBN Publishers Plc

Head Office: 1 Ighodaro Road, Jericho, P.M.B. 5205, Ibadan, Nigeria
Phone   :   02 8726701
E-mail   :   info@hebnpublishers.com
                 hebnpublishers@yahoo.com
Website   :   http://www.hebnpublishers.com

Area Offices and Branches
Abeokuta • Abuja • Akure • Bauchi • Benin City • Calabar • Enugu •
Ibadan • Ikeja • Ilorin • Jos • Kano • Katsina • Maiduguri • Makurdi •
Minna Owerri • Port Harcourt • Sokoto • Uyo • Yola • Zaria

© Ivara Ejemot Esu 2010
First Published 2010

ISBN 978 978 081 373 4

Dedication

To my beloved late father; a man of high integrity,
resourcefulness and uncommon vision

Pa Jonathan Ejemot Esu
(1905 – 1976)

Preface

Soil Characterization, Classification and Survey is a revised version of *Fundamentals of Pedology* which was published by Stirling-Horden Publishers (Nig) Ltd, Ibadan in 1999. The change in title of the book is for greater clarity among its target audience as to its content.

The book covers required lecture notes in courses such as Principles of Soil Science, Introductory Pedology, Soil Taxonomy and Conservation, Soil Characterization and Classification and Soil Survey and Land Use Planning offered in Faculties of Agriculture, Forestry and Environmental Sciences in most Universities, Polytechnics and Colleges of Education in Nigeria and many African countries.

The book is based on my personal experiences of teaching most of these courses to undergraduates and graduate students at the Kaduna Polytechnic and the National Water Resources Institute, Kaduna (1975 – 1984), Ahmadu Bello University, Zaria (1984 – 1994), University of Uyo (2007-2008), University of Calabar (1994 – date); all in northern and southern Nigeria and at the University of Ghana, Legon (2006-2007).

For those who were familiar with the earlier edition "*Fundamentals of Pedology*", this revised edition has effected the corrections and has expanded the contents of most of the chapters in light of the ever changing concepts in Pedology. A completely new chapter has also been included to deal with Laboratory Data Used in Pedological Studies and their interpretation. The chapter is meant to assist final year and postgraduate students with the interpretation of the data often generated from soil analysis and indeed the current and appropriate methods to be used in soil characterization in the laboratory.

The chapter on Principles of Soil Classification has also been tremendously expanded to include diagnostic criteria used in soil classification. A pictorial description of each of the twelve soil orders of *Soil Taxonomy* has been included as Appendix 1. Most recent revisions in the criteria used in FAO/UNESCO Soil Classification system, namely, the World Reference Base for Classification (WRB) is also dealt with extensively, among others. A detailed glossary of most of the terms used in the book has been included in Appendix 2 as well for a much easier referencing.

The book will be must useful to students of Agriculture, Geography, Environmental Protection Management and Agricultural Education in Colleges of Education, Polytechnics and Universities in Nigeria and other African countries.

Ivara Ejemot Esu, OFR
B.*Sc (Ife) M.Sc (Minnesota) Ph.D (ABU)*
Professor of Soil Science
University of Calabar, Nigeria.

Contents   *Page*

Appendices

List of Figures

List of Plates

List of Tables

1

Introduction

Pedology as a Discipline of Soil Science

Soil Science may be defined as "that science dealing with soils, as a natural resource on the surface of the earth including soil formation, classification and mapping; physical, chemical, biological and fertility properties of soils *per se*; and these properties in relation to the use and management of soils" (Soil Science Society of America, 2001).

Soil Science is a broad specialization within the field of Agriculture which is further studied under the subdisciplines of Soil Physics, Soil Chemistry and Fertility, Soil Microbiology, Soil Conservation and Soil Genesis and Classification.

Pedology is a Greek word meaning the Science of the ground or theoretical geological Soil Science (***pedon,*** ground and ***logos***, discourse, science) and it has been used both as a synonym for Soil Science (Sigmond, 1938) and another name for Soil Genesis and Classification (Vilenski, 1957). In modern Soil Science literature, Pedology connotes the combined activity of Soil Genesis, Classification and Mapping, which is a recognized discipline of Soil Science. Pedology considers the soil as a natural body and places minor emphasis on its immediate practical utilization. The Pedologist studies, examines and classifies soils as they occur in their natural environment. His findings may be as useful to highway and construction engineers as to the farmer (Brady, 1974).

What is Soil?

To a Pedologist, "Soil is a collection of natural bodies occupying portions of the earth's surface that support plants and that have properties due to the integrated effect of *climate* (temperature and precipitation) and *living organisms* (plants and animals) acting upon *parent material* (precursor of soil) as conditioned by *relief* (topography) over periods of *time*" (Soil

Survey Staff, 1951).

The five *italicized* variables in the definition above must interact in the pattern spelt out in the definition for soil to form. The variables have, therefore, come to be known as the **five soil-forming factors.**

It is pertinent to add that the definition of soil given above is from the viewpoint of a Pedologist. This is in the sense that, soil may mean different things to different people. For example, to a local farmer, soil is a medium for plant growth and a source of nutrients for his crops. Engineers view soil as to its suitability as a building material or as a support medium for roads and buildings. The engineer is trained to manipulate the soil when its properties are known, so that it becomes more stable.

When soil is viewed as a household word, it simply means dirt. As far as the housewife is concerned, soil is that substance which relates to its stickiness or tendency to cling to shoe soles and eventually dirty carpets or floors.

A geologist or mining engineer views the soil as the debris covering the rocks or minerals which he must quarry. So, as far as they are concerned, soil is a nuisance which must be removed.

Why Study Soil?

The study of soil is important for the following general reasons:

*   Soil is the foundation for the production of food, fibre and drugs. It is the mainstay of life. Knowledge of soil is very necessary in order to continue the production of these basic human necessities.
*   As the world population especially in the third world countries continues to increase, there will be need to study the soil very intensively, so as to conserve the fertile soils and produce more within a much smaller land area.
*   To shift from subsistence to mechanized farming, we need to have an adequate knowledge of our soils so as to know how best to use the soil wisely and conservatively.
*   Knowledge of soils within potential irrigation areas is essential for economic and technical reasons. The high cost of development of irrigated agriculture requires justification by assessment of the risks and benefits to be derived from the investment. Indeed, the design of the irrigation scheme itself is dependent on a detailed knowledge of soils lying within the irrigable area.
*   Soil studies are very important in land use planning activity. Even

though many decisions about land use are political, the decision as to what use a particular soil type should be put should have a technical basis, which should be dependent on a detailed knowledge of the properties of the soils within a given geographical setting.
- Soil is the domain of archaeology. The study of soils reveal information about ancient climates, geology and peoples, which are of interest to Archaeologists who seek to elucidate information related to ancient habitats, dwellings and other cultural factors. Often, "buried topsoil" holds a rich archaeological finding.
- Soil is an important part of the earth's ecosystem which comprises air, water and soil. Soil is particularly susceptible to pollution from human activities such as sewage disposal, industrial waste products, agricultural and milling wastes as well as other natural contaminants. A study of soils is useful for the identification and eventual remediation of such environmentally harmful pollutants.

Careers in Soil Science
Several career prospects exist for students of Agriculture who elect to major in Soil Science. Such prospects include:

Private Sector Farming Business:
In the private sector, opportunities exist for Soil Scientists to be engaged in medium to large scale farming business, especially in plantation agriculture such as of PAMOL Rubber Plantation in Odukpani, Cross River Estate Limited (CREL) at Uyanga both in Cross River State. Other private Plantations nation-wide include Rison Oil Palm Plantation in Rivers State, Okutipupa Oil Palm Estate in Ondo State, Akwa Palm in Akwa Ibom State and several Cocoa Plantations located in Ikom in Cross River State and South Western Nigeria. In these enterprises, the role of Soil Scientists will be in the area of site selection through soil survey procedure, soil fertility evaluation through soil testing and appropriate fertilizer recommendations based on soil test results.

Federal and State Government Ministries of Agriculture:
Soil Scientists could be very gainfully employed in the State and Federal Ministries of Agriculture, to serve as Soil Extension Agents to rural farmers particularly in organized agricultural projects such as the ADPs, UNDP and other World Bank sponsored projects involving sustainable

agriculture. Such Soil Scientists could also be involved in policy formulation and implementation for national agricultural development.

Academic and Agricultural Research Institutions:

Soil Scientists are greatly required to participate in the research programmes of University Faculties of Agriculture, Colleges of Agriculture, Colleges of Education, Polytechnics and the numerous Agricultural Research Institutes located all over the country. Their roles in these Institutions include manpower training, laboratory testing of soil samples, soil mapping and characterization studies, etc. This is a major area of contribution as Soil Scientists cannot be excluded particularly in crop based research Institutes.

State and Federal Environmental Protection Agencies:

Federal and State Environmental Protection Agencies are involved in several soil pollution and soil erosion control studies. The role of Soil Scientists in documenting, evaluating and advising on levels of pollution and erosion control through systematic mapping, soil sampling, soil analysis and data interpretation cannot be under-estimated.

Fertilizer Manufacturing Industries and Oil Companies:

Soil Scientists will readily locate job opportunities in Fertilizer Manufacturing and Blending Companies where they are involved in quality control studies, field trials and various fertilizers and environmental control monitoring. In the Oil Companies, Soil Scientists are employed in Environmental and Safety Departments as Environmental Supervisors and in Community Services Departments related to Agricultural Extension services to oil producing communities. In this regard, the role of Soil Scientists in Organizations such as NAFCON in Rivers State, Phosphate Fertilizer Company in Kaduna and other Fertilizer Blending Companies in Kano, Kaduna and Minna, etc, as well as in areas of jurisdiction of the NDDC (Niger Delta Development Commission) cannot be underestimated.

Others:

Soil Scientists especially at the First Degree level can also be involved in the teaching of Agricultural Sciences in Secondary Schools, Vocational Technical Colleges, etc. Soil Scientists can also be engaged in self-employment by establishing their own farms and properly monitoring the site selection and nutrient requirements of crops.

Revision Questions

1. What do you understand by the term Pedology?
2. Define soil from the viewpoint of an engineer, a housewife and a geologist.
3. Why is the study of soils important?
4. List the five soil forming factors. Show the sequence of interaction of these factors in the process of soil formation.
5. Outline four major career prospects for a Soil Scientist/Pedologist in Nigeria.

References

Brady, N.C. 1974. *The nature and properties of soils.* 8th Edition. Macmillan Publishing Co., Inc., New York: U.S.A. 639 pp

Sigmond, A. A. 1938. *The principles of soil science.* (Transl. from Hungarian by A. B. Yolland; edited by G. V. Jacks), T. Murby and Co. London.

Soil Science Society of America. 2001 (Ed.) *Glossary of Soil Science Terms.* Madison, Wisconsin, U.S.A. 140pp.

Soil Survey Staff. 1951. *Soil Survey Manual,* U.S. Dept. Agric. Handbook 18. U.S. Government Printing Office, Washington D.C.

Vilenski, D.G. 1957. *Soil Science* (Transl. by a Birran and Z. S. Cole) Israel Prog. For Science Trans., Jerusalem, 1960.Available U.S. Dept. Commerce, Washington D.C.

2

Soil Description Terms

Introduction

Soil is usually perceived as a body which exists in a continuum over the surface of the earth. The continuum is, however, usually broken by rock outcrops, streams, rivers and oceans.

In studying the soil, certain terms are used so as to facilitate reference to parts of it and hence its description and classification. Such terms include the soil profile, soil horizon, soil solum, the pedon, soil individual, topsoil and subsoil, mineral and organic soils.

Soil Profile

A soil profile is defined as a two dimensional, vertical cross-section made in a landscape to expose soil. It is commonly conceived as a plane at right angles to the surface. It extends from the surface of the landscape through a number of distinguishable layers of soil (horizons) to the solid bedrock which sets the limit of the profile. Plates 1 shows a vertical cross-section made in a landscape to expose a soil profile in a road cut along Owerri – Onitsha road in Imo State, Nigeria. Plates 2 and 3 also show typically deep soil profiles while Plate 4 shows a shallow soil profile to the underlying bedrock within 50cm from the soil surface. Note that in Plate 3 the bedrock is not even exposed, while in Plate 2 the depth to weathered bedrock (parent material) is several metres from the soil surface.

In general, depending upon the geographical location and exposure time, different profiles have different horizons and thicknesses. The horizons may be of many kinds, such as layers with organic matter accumulation in the soil surface, accumulations of clays in subsoil layers, deposits of salts or carbonates in subsoils or layer of cementation (hardpans) (Donahue *et al.*, 1983).

Plate 1: *A vertical cross-section made on a landscape to expose soil along a road-cut (soil profile) in Imo State, Nigeria*

Plate 2: *A deep soil profile on coastal plain sands with the parent material exposed only at great depth at Owerri, Nigeria*

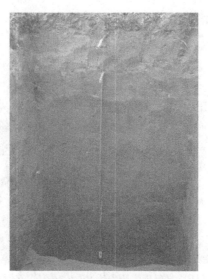

Plate 3: *A deep soil profile with the underlying bedrock several metres below the soil surface at Ahoada, Rivers State, Nigeria*

Plate 4: *A shallow soil profile in a road-cut exposing   an underlying granitic bedrock at about 50cm depth along Ogoja-Odajie Mbube road in Cross River State, Nigeria*

Soil Horizon

A soil horizon is a layer of soil or soil material approximately parallel to the land surface and differing from adjacent genetically related layers in physical, chemical and biological properties or characteristics such as colour, structure, texture, consistency, kinds and numbers of organisms present, degree of acidity or alkalinity, etc. In other words, a horizon is a unit of a soil profile having its own specific characteristic, such as can be seen as pegged in Plate 3.

Processes which tend to promote the development of horizons within a given soil profile as illustrated in Fig. 2.1 include:

a. Additions to the soil of organic matter, erosional sediments, water in form of precipitation and energy from the sun.
b. Transformation processes such as the weathering of rocks and minerals, humification of organic matter, as well as clay and organic matter reactions.
c. Translocation processes which involve the movement within soils of clay, organic matter and sesquioxides (Fe and Al oxides) through eluviation-illuviation processes. Other translocated materials like soluble salts in water and the circulation of nutrients by plants.
d. Losses of materials such as water by evapotranspiration, nutrient losses by leaching and crop harvest, nitrogen by denitrification, soil enmasse by erosion and energy by radiation.

The major soil horizons within any given hypothetical soil profile as shown in Figure 2.2 are designated by letters O, A, E, B, C, and the underlying bedrock designated R. The horizons have these broad characteristics:

O Horizon - Horizon which is formed mainly from organic litter derived from plants and animals. Such a horizon is dominated by fresh or partly decomposed organic materials.

A Horizon - Horizon characterized by dark colour which is formed at or adjacent to the soil surface and characterized by the accumulation of decomposed organic matter as well as resistant minerals of sand and silt size.

E Horizon - Horizon in which the main feature is loss of silicate clay,

Fe, Al or some combination of these leaving a concentration of sand and silt particles such that the layer appear somewhat bleached. It is a zone of major eluviation. In old textbooks, this horizon would be designated as A2.

B Horizon -   Horizon characterized by the accumulation of colloidal materials notably clay, Fe and Al oxides and silicate minerals which are eluviated from the overlying horizons. And also through the weathering of parent material from the C-horizon.

C Horizon -   Horizon of partially weathered material from which the solum is presumed to have formed (parent material). It generally underlies the B horizon and is presumed to have been differentiated from the underlying bedrock.

R Layer   -   Hard bedrock, such as granite, basalt, quartzite or sandstone, etc.

Additions to the soil

Water as precipitation, condensation or runon

O_2 and CO_2 from the atmosphere

N, Cl, and S from atmosphere and precipitation

Organic matter from biotic activity

Material as sediments

Energy from the sun

Translocation within

Clay, organic mater and sesquioxides by water

Nutrients circulated by plants

Soluble salts in water

Soil enmasse by animals

Losses from the soil

Water and materials in solution or suspension

Losses from the soil

Water by evapotranspiration

N by denitrification

C as CO_2 from oxidation of organic mater

Soil enmasse by erosion

Energy by radiation

Transformations within

Humification of organic matter

Reduced particle size by weathering

Structure and concretion formation

Mineral transformations by weathering

Clay and organic matter reactions

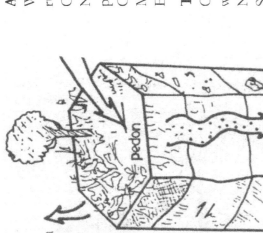

Fig. 2.1: Processes which promote the development of horizons in soils

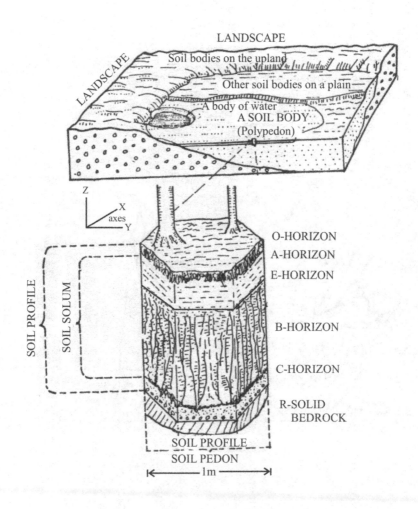

Fig. 2.2: *A pedon of a soil body within a landscape*

Soil Solum

The soil solum is an incomplete soil profile that may be simply defined as "the genetic soil developed by soil-building forces" (Soil Survey Staff, 1962). It is generally considered as that part of the soil profile which is influenced by plant roots. It includes the A, E and B horizons which are often referred to as the "true soil" and excludes the parent material of the O or C horizons as indicated in Figure 2.2.

The Pedon

A pedon is the smallest volume that can be called "a soil" (Soil Survey Staff, 1960). It is obvious that the whole soil or even areas of it cannot be studied at one time. Therefore, a small three-dimensional, roughly hexagonal shaped sample of the soil which is termed the pedon is used.

A pedon is, therefore, the smallest volume that we should describe and sample to represent the nature and arrangement of horizons of a soil profile and variability in its other properties that are preserved in samples. It is thus, comparable in some ways to the unit cell of a crystal. The lower limit of the pedon is the somewhat vague limit between the soil and the "not-soil" or bedrock below. The lateral dimensions are large enough to represent the nature of any horizon which may be seen in a two-dimensional soil profile, when viewed in a pedon. The area of a pedon ranges from 1 to 10 square metres, depending on the variability of the soil body. Figure 2.2 illustrates the concept of a pedon, and indicates that the pedon merely enlarges upon the concept of the soil profile to include a lateral as well as a vertical extent of a soil and puts limits on the volume to be considered.

Polypedon/Soil Individual/Soil Body

Soil as a three-dimensional component of the landscape, is located in space along a vertical axis between the free air above and the geologic substratum (bedrock) below and in a horizontal plane to material that is not considered soil such as deep water, rock outcrops, buildings, etc. similarly, an individual soil body or soil taxonomic unit consists of contiguous similar pedons that are bounded on all sides by either "not-soil" or by pedons of unlike character. This group of contiguous similar pedons is called a *polypedon* or a *soil body* or a *soil individual*. The limits of the soil individual are also the conceptual limits between soil series, which are the classes of the lowest category in *Soil Taxonomy* (Soil Survey Staff, 1975). In fact, the practice of arranging soils into taxonomic units derives from the concept of the soil individual or soil body or polypedon. Polypedons link the real bodies of soil in nature to the mental concepts of taxonomic classes (Soil Survey Division Staff, 1993).

Sequum/Bisequal/Buried Soil

A *Sequum* (plural, *Sequa*) is a B horizon together with any overlying E (eluvial) horizons. A single sequum is considered the product of a specific combination of soil-forming processes (eluviation-illuviation). Most soils have a single sequum, but some have two or more Sequa. Soils in which two Sequa have formed one above the other in the same deposit is said to be *Bisequal*. If two Sequa formed in different deposits at different times, the soil is not Bisequal but a *Buried soil*.

Topsoil and Subsoil

In any given soil body or profile, the topsoil also called surface soil and the subsoil also termed subsurface soil are generally recognized.

The topsoil usually consists of the upper 0 – 30cm depth or the A-horizon of the soil profile. It is generally dark coloured with high organic matter and plant nutrient contents. It is thus the major zone of root development. The topsoil is the layer which is often ploughed and cultivated and is therefore subject to manipulation and management. For instance, its fertility and to a lesser degree its productivity may be raised, lowered or satisfactorily stabilized at levels consistent with economic crop production. Also, the depth of the topsoil may vary depending on the extent of soil loss through erosion.

The subsoil like the name implies, is the zone below the topsoil and is usually synonymous with the B and C horizons of the soil profile. The subsoil is extremely important both as an anchorage for the roots of plants and as the supplier of reserved soil moisture to plants. Therefore, the productivity of a soil is determined to a large degree by the nature of its subsoil. This is why parameters such as depth to physical or chemical barriers (effective soil depth), subsoil texture and drainage conditions are very important in the evaluation of the productivity of a soil.

Mineral and Organic Soils

A mineral soil is a soil consisting predominantly of, and having its properties determined predominantly by mineral matter. Such mineral matter is formed by the dissolving and disintegration of rocks. Usually, mineral soil contains less than 15% organic matter but may contain an organic surface layer up to 30cm thick. Mineral soils occupy a very high proportion of the total land area of soils on earth and are consequently of greater importance than organic soils.

An organic soil on the other hand, is a soil which contains a high percentage, generally greater than 15% of organic matter throughout the solum. The high accumulation of organic matter in these soils is often due to low oxygen environment of shallow and stagnant waters. Organic soils are variously called peat, bog or muck soils and are generally common in the temperate regions of the world where extremely cold winter temperatures inhibit organic matter mineralization. Countries or regions which have particularly large reserves of peat soils include Finland, Canada, Northern U.S.A., Ireland and Alaska.

Revision Questions

1. With the aid of an annotated diagram, show the differences between a pedon, a soil profile and soil solum.
2. Outline the processes which tend to promote the development of classical soil horizons in any given soil profile.
3. Distinguish clearly between the following pairs of concepts:
    (a) Pedon and Soil Individual
    (b) A Horizon and C Horizon
    (c) Topsoil and Subsoil
    (d) Mineral Soil and Organic Soil
    (e) A Bisequal and a Buried Soil

References

Donahue, R. L., R. W. Miller and J. C. Shickluna, 1983. Soils: an introduction to soils and plant growth. Fifth edition. Prentice-Hall Inc., Englewood Cliffs, New Jersey, 667pp.

Soil Survey Staff, 1960. Soil classification, a comprehensive system – 7th approximations. U.S. Dept. Agric, U.S. Govt. Printing Office, Washington D.C.

Soil Survey Staff, 1962. Supplement to Agriculture Handbook 18, Soil Survey Manual (replacing pages 173-188) U.S. Dept. Agric. U.S., Govt. Printing Office, Washington D.C.

Soil Survey Staff. 1975. Soil Taxonomy. A Basic System of Soil Classification for Making and Interpreting Soil Surveys. Agric. Handbook No. 436. U.S. Govt. Printing Office, Washington D.C. 754 pp.

Soil Survey Division Staff. 1993. Soil Survey Manual. SCS-USDA Handbook 18 (modified 2006). 279pp.

3

Components of Soil

Introduction

Mineral soils consist of four major components:

a.   Mineral matter
b.   Organic matter
c.   Soil water
d.   Soil air

These components exist mostly in a fine state of subdivision, such that they are so intimately mixed and satisfactory separation is rather difficult (Brady, 1974).

It is pertinent to note that the percentage contribution of each of the components enumerated above to the total volume of soil would vary from the topsoil to the subsoil and would also depend on the type of soil. For example, in a soil which was described as a "silt loam surface soil in optimum condition for plant growth", Brady (1974) indicated that the composition of the various components by volume was as follows:

Mineral Matter .... 45% by volume
Organic Matter .... 5% by volume
Soil Water .... 25% by volume
Soil Air .... 25 % by volume

This can be represented pictorially as shown in Fig. 3.1.

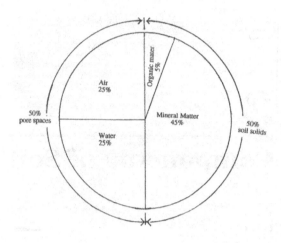

Fig. 3.1: *Volume Composition of a silt loam surface soil in optimum condition for plant growth.*

It is pertinent to note that the proportion of air and water in the soil is subject to great fluctuations under natural conditions, depending on the weather and other factors. The volume composition of subsoils is also somewhat different from those of topsoils. For instance, subsoils usually contain lower amounts of organic matter, and are usually more compact and often contain higher percentages of micropores. This implies that they have higher percentages of mineral particles and water than the topsoils.

In general though, a soil is composed of 50% solid materials and 50% of pore spaces, but each of the four components of a particular soil will vary depending on its texture, climatic setting, whether the sample is obtained from topsoil or subsoil and the treatment to which it has been subjected.

Mineral Matter

The mineral matter component of soil is variable in size and composition. It includes rock fragments and minerals of various kinds, which are remnants of massive rocks from which the parent material (regolith) and in turn the soil solum have been formed through the process of weathering. The rock fragments range in size from boulders to clay-size particles of the order of 0.002mm diameter.

Mineral matter is said to be primary if it is derived from solid massive rocks with little change in composition from the original country rock. It is termed secondary if it has been formed by the weathering of less resistant minerals. In general, the primary minerals are most prominent in the finer materials, especially in clays. Clearly, mineral particles have much to do with the properties of soils in the field.

Some of the important functions of mineral matter in soils include:

i.   Mineral materials form the body of the soil and so give the soil its mass.

ii.   Mineral materials give mechanical support to plants, buildings, dams, etc.

iii. They hold and supply nutrients to the plants that they support,

iv.   Soil composition of mineral materials determines to a large extent the use to which a particular soil can be put.

Organic Matter

Organic matter may be pictured as an assemblage of the following:

i.   Fresh fallen leaves, twigs, stems, etc.

ii.   Roots of plants.

iii. Partly decomposed products of (i) and (ii) above.

iv.   Humus, which is the decomposed products of (i) and (ii) above.

Humus is usually black or brown in colour and is colloidal in nature. Its capacity to hold water and nutrient ions greatly exceeds that of clay, its inorganic counterpart. Small amounts of humus thus augment remarkably the soil's capacity to promote plant growth. Humus is indeed the colouring matter that we see in horizons of surface soils.

Organic matter is often said to exist in a dynamic state in soil. This is in the sense that it is constantly being added and at the same time being broken down by soil microorganisms to humus. Organic matter is also subject to eluviation from topsoil horizons. It is involved in several nutrient cycle reactions, e.g. the Nitrogen, Sulphur and Phosphorus Cycles. This is why organic matter hardly accumulates in soil to a large extent, except under unfavourable decomposition conditions such as under anaerobic or/and very cold environments.

Some functions of the organic matter component of soils are as follows:

i.   Organic matter is a major source of plant nutrients. The main essential nutrients supplied by organic matter include nitrogen, phosphorus and sulphur as well as trace elements (Cu, Mo, Zn, Mn, Fe, B, and Cl).

ii.   Organic matter is a soil binding agent; it thus promotes the development of good soil structure.
iii. Through its granulating effect on the physical condition of soils, organic matter also increases the water holding capacity of soils.
iv.   Organic matter is the main source of energy for soil microorganisms; it, therefore, encourages the biological and biochemical processes operating in the soil.

Soil Water

Water and air competes for the pore spaces in any soil. Water usually occupies the pore spaces that are not occupied by air. Soil water is usually held around mineral particles or grains occupying the large pores between the mineral grains.

Soil water is often held by varying degrees of forces around mineral particles and based on the amount of force holding the water; we can classify soil water into three categories as follows:

i.   *Hygroscopic Water,* which is water held very tenaciously as films on mineral particles, at suction values greater than 31 bars. It is equivalent to water remaining in the soil after "air-drying". It is essentially non-liquid and moves primarily in the vapour form. Higher plants cannot absorb hygroscopic water, but some microbial activity has been found to take place in soils containing only hygroscopic water (Brady, 1974).

ii.   *Capillary Water,* which is water held less tenaciously than hygroscopic water. The water is held in the capillary or small pores of a soil and behaves according to laws governing capillarity. Such water includes most of the water taken up by growing plants and exerts suctions between 0.1 and 15 bars.

iii. *Gravitational Water,* which is excess water which has not been drained. It may be described as soil water in excess of the field capacity (0.1 to 0.3 bars). Gravitational water is of little use to plants because it occupies the larger pores, thereby reducing soil aeration. Its removal from the soil in drainage is generally a requisite for optimum plant growth. It occupies the larger soil pores and moves readily under the force of gravity.

The classifications of soil water outlined above indicate that not all the water in the soil is available to the plants for performing their growth function. Much of it remains in the soil as unavailable water even though plants growing on such a soil will be seen to be wilting as a consequence of water shortage.

Soil water has the following functions:

i.   Water is a universal solvent which plays a leading role in the weathering of soil minerals.

ii.   Soil water together with its dissolved salts makes up the soil solution which is so important as a medium for supplying nutrients to growing plants.

iii. It affects the physical properties of soil such as drainage, colour, structure, consistence, tilth, etc.

Soil Air

The content and composition of soil air is determined to a large degree by the soil-water relationships. Soil air simply moves into those soil pores not occupied by water. Following flood irrigation or heavy rainfall, large pores are the first to be vacated by soil water, followed by medium-sized pores as water is progressively removed by evaporation and plant utilization. Thus, soil air ordinarily occupies the large-size pores, also termed Macropores and as the soil dries out, the intermediate sized pores termed Mesopores. This explains the tendency for soils with a high proportion of tiny pores or Micropores, to be poorly aerated. In such soils, water dominates and the soil-air content and composition is unsatisfactory for optimum plant growth.

Soil air differs from atmospheric air in the following respects:

i.   Soil air exists in a sort of discontinuous phase, usually located in the maze or matric of soil pores, separated by soil solids. This fact accounts for its variation in composition from place to place in the soil. In local pockets, reactions involving the gases can greatly modify the composition of soil air.

ii.   Soil air generally has a higher moisture content than atmospheric air; the relative humidity of soil air approaches 100% when the soil moisture is optimum.

iii. The content of carbon dioxide is often several hundred times more concentrated than the 0.03% commonly found in the atmosphere.

iv.   Oxygen content decreases accordingly, and in extreme cases may be no more than 10 to 12% as compared to about 20% for normal atmospheric air.

Pedologically, soil air is important from the point of view of the following functions:

(a)   A soil profile that is well aerated favours the activities of soil micro-organisms and so promote plant growth.

(b)   Good aeration makes soil forming processes to operate faster.

Revision Questions

1.   Make a pictorial representation of the volume composition of a *named* mineral soil.

2.   "In general, a soil is composed of 50% solid materials and 50% of pore spaces." Discuss this view.

3.   What do you understand by the term Organic matter? What are its functions in soil?

4.   List four functions of mineral materials in soil.

5.   Gravitational water is the most easily lost from soils following precipitation. Discuss this view vis-à-vis the other types of soil water.

6.   Distinguish between Macropores, Mesopores and Micropores in soils. In what ways does soil air differ from atmospheric air?

Reference

Brady, N. C. 1974. *The nature and properties of Soils.* 8th Edition. Macmillan Publishing Co., Inc., New York: USA. 639 pp.

4

Soil Forming Rocks and Minerals

Introduction

Soil particles are the discrete units which comprise the solid phase of the soil. These particles cluster together as aggregates or peds, but can be separated from one another by chemical and mechanical means. The particles have diverse composition and structure and generally differ from one another in both size and shape. They may be organic or inorganic, crystalline or amorphous.

In this chapter, we shall be concerned mainly with the origin and properties of the inorganic fractions of soil which are derived from rocks which in turn consist of different types of mineral species.

Soil Forming Rocks

A Rock may be defined as a considerable mass of two or more minerals. Rocks are classified into three groups according to their mode of formation as Igneous, Sedimentary and Metamorphic Rocks.

Igneous Rocks

These are rocks formed by the cooling and subsequent solidification of molten material called magma. The kind of igneous rock formed depends on (a) the composition of the magma and (b) the rate at which the magma cooled.

Thus, a magma containing 60% or more silica (i.e. silicon and oxygen existing either in the free form as quartz (SiO_2) or in combination with other elements as silicates) solidifies to form igneous rocks referred to as acidic igneous rocks. Those containing less than 50% silica produce basic igneous rocks while magmas consisting of 50-60% silica give rise to intermediate igneous rocks. The rate of cooling, on the other hand, determines the texture of the igneous rocks. For example, if cooling is slow, as in the case of magma within the earth's crust, the igneous rock would be coarse textured because under these conditions of slow crystallization, the mineral particles would have an opportunity to grow to considerable size. Such rocks are commonly called *plutonic* or *intrusive* rocks. However, where magma cools suddenly, as in the case where it is extruded onto the earth's surface by volcanic action, the igneous rock will be fine grained and such rocks are known as *extrusive* or *volcanic* rocks.

By combining the names derived from these two variables (reaction and texture), it is possible to classify igneous rocks into major categories with examples as shown in Table 4.1.

Table 4.1:   *A Simplified classification of common igneous rocks according to the size of mineral grains in the rock and reaction*

Rock Texture	Reaction		
	Acidic	**Intermediate**	**Basic**
Coarse grained (Intrusive)	Granite	Diorite	Gabbro
Medium to fine (Extrusive)	Rhyolite	Andesite	Basalt
Fine to very fine (Glassy)	Obsidian (Dense) - Felsite Pumice & Scoria (Vesicular)		Basalt Glass

Sedimentary Rocks

These are rocks formed from sediments deposited in a body of water or by wind. The sediments are held together by various cementing agents under the influence of weak pressure. The materials from which sedimentary rocks are derived, are pre-existing rocks first disintegrated through physical and/or chemical weathering processes. They are thus said to be secondary in origin.

Sedimentary rocks can be classified on the basis of their mode of formation and the grain size as follows:

(a) Mode of formation

(i)   Rocks with distinct visible grains of pre-existing rocks often termed *clastic rocks* e.g. conglomerates, sandstones;

(ii)   Rocks precipitated from pre-existing rocks which are generally of clay-size; termed chemical sedimentary rocks e.g. gypsum, limestone, chert (SiO_2) and flint (glassy).

(iii)   Rocks resulting from pressure of organic deposits e.g. coal and calcareous reef.

(b) Basis of grain size

(i)   Coarse grained rocks e.g. conglomerate and breccia which is made up of various sized fragments of rocks cemented together.

(ii)   Medium grained rocks e.g. sandstones which are mainly cemented sands (quartz and feldspars).

(iii)   Fine grained rocks e.g. shales, which are cemented clays and silts and limestones which are cemented calcium carbonate, clays, silts and fine sands. Dolomites are similar to limestones, but have more magnesium carbonates as part of the carbonate mixture.

Metamorphic Rocks

These are rocks which have been formed by marked transformation of igneous and sedimentary rocks under the influence of heat and/or pressure.

Metamorphic rocks are distinguished from their sedimentary counterparts by their crystalline structure and from their igneous counterparts by a banded or laminated arrangement of their minerals. The grains in the metamorphic rocks tend to be aligned perpendicular to the line of stress. The banded materials in metamorphic rocks unlike in sedimentary rocks are not homogeneous hence one might see bands of white, black or brown lines.

Common examples of metamorphic rocks include the following:

(i)   *Gneiss* (pronounced "nice"): Formed mostly from granites, rhyolites, andesites and other similar mocks. The minerals in gneiss are oriented to form light and dark bands.

(ii)   *Schist:* Layered or flaky rock composed mainly of micas or biotite e.g. (mica schist, biotite schist, granite schist).

(iii) *Slate:* This is mainly hardened shale or siltstone.

(iv)   *Quartzite:* This is mainly recrystallized quartzitic sandstone, formed by heat and/or pressure. It is very slow to weather, producing generally sandy shallow soils.

(v)   *Marble:* Derived from limestone or dolomite which are hardened enough to polish.

Rock Forming Minerals

A mineral is a naturally occurring inorganic compound which has a fairly defined natural composition and definite physical, chemical and crystalline properties. Minerals are classified according to their origin and chemical composition.

Based on origin, minerals may be primary or secondary. Primary minerals are formed by the cooling and solidification of molten material while secondary minerals originate from the least resistant primary minerals as products of chemical weathering.

Common examples of primary and secondary minerals are shown in Table 4.2.

In general, rock forming minerals can be classified into six main groups as follows:

Silicate Minerals

These are by far the most important group of minerals in soils. The major elements of silicate minerals are Si, Al, O and H which exist in various coordination to form many common rock forming minerals in soils. Common minerals which belong to this group include quartz, feldspars, micas, olivine, zircon, pyroxenes, amphiboles, tourmaline and the clay minerals.

Table 4.2: *Selected common primary and secondary minerals in soils*

Name	Formula	Comments
	Primary Minerals	
Quartz	$Si O_2$	Hard, very resistant to weathering, major constituents of sands.
Feldspars:		Termed potash feldspars, weather
Orthoclase)	$(K, Na) AlSi_3O_8$	slowly to moderately, major source
Microcline)		of K in soils.
Na plagioclase	$NaAlSi_3O_8$	In the continuous series of plagioclases
Ca plagioclase	$CaAl_2Si_2O_8$	$(Ca, Na) Al (Al, Si) Si_2O_8$
Micas:		
Muscovite	$KAl_3Si_3O_{10} (OH)_2$	Flaky and glitters in rocks and wet sands; resistant to weathering
Biotite	$KAl (Mg, Fe)_3Si_3O_{10} (OH)_2$	Important source of K and clay.
Amphiboles:		
Hornblende	$Ca_2Al_2Mg_2Fe_{31}Si_6O_{22} (OH)_2$	Included in a group of dark minerals along with biotite called "ferromagnesian
Augite	$Ca, (Al, Fe) (Mg, Fe)_4Si_6O_{24})$	minerals", weathers
Hypersthene	$(Mg, Fe)_2 (Si_2O_6)$	fast relatively and forms clay.
Apatite	$(3Ca_3 (PO_4)_2CaF_2$	Most common mineral supplying phosphorus
Tourmaline	$Na (Mg, Fe)_3Al_6BO_3Si_6O$	Resistant mineral to weathering; main source of B in soils.
	Secondary Minerals	
Calcite	$CaCO_3$	Slightly soluble materials in limestone or dolomite rock, common in arid region soils;
Dolomite	$CaMg (CO_3)_2$	plant sources of Ca and Mg.
Gypsum	$CaSO_42H_2O$	Soft, moderately soluble mineral Found in arid regions.
Iron Oxides:		
Hematite	Fe_2O_3	A group of minerals present in
Limonite	$Fe_2O_33H_2O$	highly weathered soils giving
Goethite	$FeOOH$	soils yellow to reddish colours;
Magnetite	Fe_3O_4	source of nutrient Fe.

Aluminium Oxides:

Gibbsite	$Al_2O_3 3H_2O$	A group of minerals present in
Bayerite	$Al(OH)_3$	highly weathered, highly leached
Boehmite	$AlOOH$	soils.
Corundum	Al_2O_3	

Clay Minerals:

Kaolinite	$Al_2Si_2O_5(OH)_4$	Kaolinite, illite and chlorite are
Illite	$KAl_2(Si_3Al)O_2 10(OH)_2$	non-expanding minerals, found
Chlorites	$(AlMg_2(OH)_6)_8 Mg_3$	mostly in subhumid to humid,
	$(Si_{4-x}Al)O_2 10(OH)$	leached soils, while
Vermiculites	$(Mg(H_2O)_6)_2 n(Mg,Fe_3)$	vermiculite, smectites
		(montmorillonite) are
	$(Si_{4-n}Al_n O_{10}(OH)_2$	expanding clay minerals found
		in arid to humid soils which are less
		leached.
Montmorillonite	$KAlSi_{11}O_2 30(OH)_6$	
Smectite	$Nax[Ca_{12-x}Mgx) Si_4O_{10}(OH)_2]$	

Iron Oxide Group

These minerals are usually formed as a result of intense chemical weathering and are most common in the tropics. Common examples include hematite (Fe_2O_3), which is responsible for the reddish colouration in many tropical soils and goethite ($FeOOH$) which imparts a dark yellowish-brown colour to soils. Other examples include magnetite (Fe_3O_4), which is a magnetic iron oxide inherited from the parent rock and limonite($Fe_2O_3.3H_2O$) which is a reduced form of hematite which imparts yellow colour to soils.

Carbonate Group

The minerals in this group are important in soils for the calcium and magnesium that they contain. Carbonates are commonly found in arid regions containing limestones and marble. Soil carbonate minerals like magnesite, dolomite and calcite are known to adsorb phosphates and trace metal elements, especially in soils with alkaline reaction.

Phosphate Group

Apatite ($Ca_5(F,Cl)(PO_4)_3$) is a major mineral of this group. It is an accessory mineral in slightly to moderately weathered soils. It is the native source of most soil phosphorus. It may be inherited from igneous and metamorphic rocks as well as from some limestones (Allen & Fanning, 1983).

Sulphate Group

Gypsum ($CaSO_4.2H_2O$) is the most important mineral of this group and occurs mainly in soils of arid regions. Jarosite ($KFe_3(SO_4)_2(OH)_6$), formed as a result of oxidation of sulphides, has been identified in tidal marsh soils ("cat clays") after drainage; gypsum may also form if sufficient Ca is present. Other minerals of this group include barite ($BaSO_4$) epsomite ($MgSO_4.7H_2O$) thenardite (Na_2SO_4) and Pyrite (FeS_2), also found in tidal marsh.

Other Minerals

These include halide/chlorides e.g. halite (Na Cl), fluorides, rutile (TiO_2) and ilmenite ($Fe\ TiO_3$).

Weathering of Rocks and Minerals

Weathering refers to the chemical and physical disintegration and decomposition of rocks and minerals which are not at equilibrium under the temperature, pressure and moisture conditions of the atmosphere-lithosphere interphase.

The geological work accomplished by weathering is of two kinds:

(a) physical or mechanical changes in which minerals are disintegrated by temperature/pressure changes, frost action, erosion and organisms degrading into medium and fine sized particles.

Rock $\xrightarrow{\text{Disintegration}}$ gravel, sand, silt particles (increase in surface area)

(b) chemical changes in which minerals are decomposed, dissolved and loosened by the H_2O, O_2 and CO_2 of the atmosphere and by soil water, organisms and the products of their decay.

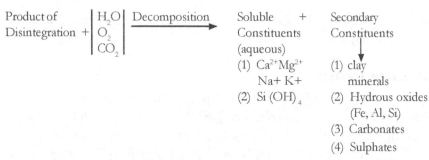

   The physical, chemical and biological agents actively co-operate with one another in the process.

Why Study Weathering?

Weathering is studied for the following reasons:

(i)   It initiates the formation or genesis of mineral soils by producing the parent material from which soils are formed.

(ii) Weathering results in the formation of secondary clay-sized silicate minerals which are responsible for holding cations such as Na^+, K^+, Mg^{2+}, Ca^{2+}, Zn^{2+}, Fe^{2+}, etc. at their exchange sites and releasing these ions to plants. They also hold water especially in the colloidal fractions.

(iii) It is the means by which plant nutrient elements are derived and replenished. In general, the central concern of soil scientists is the weathering of minerals and the release of these elements as ions for plant use.

(iv) In soil, the degree of weathering as assessed by the presence or absence of weatherable minerals indicates the stage of soil development from which the fertility or infertility of the soil can be assessed.

Physical Processes of Weathering

Physical processes involved in weathering always results in the reduction in size or disintegration of rocks and minerals and thereby provide a larger surface area which can be more easily decomposed or weathered chemically. Note that disintegration only results in a decrease in size of rocks and minerals without appreciably affecting their chemical composition. While during the process of decomposition (chemical weathering), definite chemical changes take place, soluble materials are released and new minerals are synthesized or left as resistant end products.

Some of the processes involved in physical weathering may be outlined as follows:

(a) *Thermal Expansion and Contraction*

Disintegration of rocks and minerals often occurs during heating and cooling. This is in the sense that rocks are aggregates of minerals which differ in their coefficients of expansion upon being heated. With every temperature change, therefore, differential stresses are set up which eventually must produce cracks and rifts thus encouraging mechanical breakdown.

Also, as a rock is heated by exposure to the sun during the day, the surface of the rock becomes warmer than the interior of the rock. Due to the higher temperature, the expansion of the surface is greater than the interior of the rock. When the sun goes down, the opposite is true, and the rock losses heat by radiation to the atmosphere. Under these conditions, the surface of the rock becomes cooler than the interior and consequently contracts to a greater degree. This differential expansion and contraction causes the rock to develop planes of weakness and thin sheets of the rock tend to "flake off". This flaking off is called exfoliation and indicates that the rock is weathering physically.

(b) *Grinding Action of Particles*

The erosive action of water, wind and gravity causes small pieces of rocks and soil particles to be carried along striking other rocks and leading to their disintegration.

(c) *Freezing and Thawing*

The expansion force of water as it freezes and thaws in cold climates is sufficient to split minerals or rocks. Freezing water can exert a pressure of $146kg.cm^3$. This phenomenon is common in the cold temperate regions of the world.

(d) *Action of Organisms*

When simple plants like mosses and lichens grow upon the exposed surfaces of rocks, they secrete acidic substances which assist in the dissolution of minerals within rocks. Also, growing plant roots enter crevices of rocks and are capable of splitting many rocks. Similarly, animal burrowings, ploughing and cultivation by man and other similar processes accelerate the process of physical disintegration of rocks and soil minerals.

Chemical Processes of Weathering

Chemical processes of weathering are by far the most important reactions which result in the decomposition of rocks and minerals in soils. The processes result in the dissolution of minerals through solubility and structural changes.

Solubility changes are caused by solution (usually in water), hydrolysis, carbonation and oxidation-reduction reactions, while structural changes in rocks and minerals are brought about usually by the processes of hydration and oxidation-reduction. The processes are further discussed as follows:

(a) Solution

Water is important in altering the chemical properties of rocks by the process called solution. Water moving over or through rocks dissolve and carry with it soluble portions of the rock. In addition, the water carries along the soluble products of decomposition such as KOH and Ca $(OH)_2$. The effectiveness of water in decomposing minerals is enhanced by the presence of salts, bases and acids.

The process of solution may be illustrated by the dissolving of calcium carbonate contained in a loess deposit in water and the dissolving of salt in water as follows:

$$Ca\,CO_3(s) + H_2O + CO_2\,(g) \longrightarrow Ca^{2+}\,(aq) + 2HCO_3^-\,(aq)$$

$$Na\,Cl_{(s)} + H_2O \longrightarrow Na^+, Cl^-, H_2O$$
Dissolved ions, surrounded by water molecules.

(b) Hydrolysis

When the water molecule (HOH) splits into H^+ and OH^- and each part combines with other elements, hydrolysis is said to have occurred. The new substances formed are usually more soluble than the original material. It is one of the most important weathering processes causing soil profile changes. The H^+ in water tends to replace the positively charged cation in many compounds especially when the compound contains strongly basic elements. The OH⁻ radical, on the other hand, combines with the positively charged cation to form hydroxides and other new substances.

The hydrolysis of silicate or aluminosilicate minerals is very common in soils and may be represented by the following chemical equations:

$$KAlSi_3O_8 \;+\; HOH \longrightarrow HAlSi_3O_8 \;+\; KOH$$

Microcline		(Acid silicate Clay)	(Potassium) hydroxide)

Very slowly soluble   (more soluble)   (very soluble)

$$2HAlSi_3O_8 + 8HOH \longrightarrow Al_2O_3.3H_2O + 6H_2SiO_3$$
                                   Gibbsite

The potassium released by the first reaction is soluble and can be adsorbed by soil colloids, taken up by plants as a nutrient or leached out of the soil system. The aluminium and silicon compounds in the second equation may recrystallize into a clay mineral such as kaolinite.

(c) *Carbonation*

Carbonation is the reaction of a compound with carbonic acid (H_2CO_3), a weak acid produced when gaseous carbon dioxide is dissolved in water.

The respiration of plant roots and decomposition of organic matter results in an enrichment of the medium with CO_2. The carbon dioxide combines with water to produce carbonic acid which is more effective in dissolving minerals than pure water. An example is the chemical solution of calcite, a slightly soluble mineral, into more soluble bicarbonates as illustrated in the equation below:

$$Ca\,CO_3 \;+\; H_2CO_3 \longrightarrow Ca(HCO_3)_2$$

(Calcite: only slightly   (Carbonic acid)   (Calcium
Soluble)   bicarbonate:
                                                    Readily soluble)

(d) *Hydration*

Hydration is the combination of a solid chemical such as a mineral or a salt with water often without actual decomposition or modification of the mineral itself. However, the hydration water combining with the mineral, changes the mineral structure by increasing its volume and thereby making it softer, as micas become hydrated, some H^+ and OH^- ions move in between the platelike layers. In so doing, they tend to expand the crystal and make it more porous, thus hastening other decomposition processes (Brady, 1974).

Another good example of hydration is the development of limonite from hematite, which is shown as follows:

$$2Fe_2O_3 \ + \ 3H_2O \longrightarrow 2Fe_2O_3.3H_2O$$

Hematite   Limonite

(red)   (yellow)

When the products of hydration dry out because of varying weather conditions, dehydration may occur. Thus, limonite may readily be changed back to hematite with a noticeable change in colour.

(e) *Oxidation - Reduction*

Oxidation is both the chemical combining of oxygen with a compound and the loss of electrons (change in oxidation number) of some chemical element. Oxidized minerals have a volume increase with added O_2 and are usually softer than the unoxidized material. If an element's oxidation number is changed, this can also imbalance the mineral's electrical neutrality, making it more easily attacked by water and carbonic acid.

Oxidation is particularly manifest in rocks containing iron, and element which is easily oxidized. In some minerals, iron is present in reduced ferrous (Fe++) form. If oxidation to the ferric ion takes place while iron is still part of the crystal lattice, other ionic adjustments must be made since a three-valent ion is replacing a two-valent one. These adjustments result in a less stable crystal which is then subject to both disintegration and decomposition. In other cases, ferrous ion may be released from the crystal and is almost simultaneously oxidized to the ferric form. An example of this is the hydration of olivine and the release of ferrous oxide, which may also be immediately oxidized to ferric oxide (hematite):

$$3MgFeSiO_4 \ + \ 2H_2O \longrightarrow H_4Mg_3Si_2O_9 + SiO_2 \ + \ 3FeO$$

Olivine   Serpentine   Ferrous

                  Oxide

$$4FeO + O_2 \longrightarrow 2Fe_2O_3$$

Ferrous Oxide   Hematite

Reduction is the reverse of the process of oxidation, it involves the chemical process in which electrons are gained, the negative charge is increased, and the positive charge is decreased. In soils, reduction usually occurs when oxygen is scarce, as in stagnant water conditions. Reduction in minerals may result in electrically unstable compounds, in more soluble ones, or in changes in atom size causing internally stressed conditions (Donahue *et al.*, 1983). All these eventually cause faster decomposition of minerals and rocks.

Factors Governing the Stability of Minerals to Weathering

It is generally observed that different minerals have relatively different degrees of stability to weathering. For instance, while the mineral olivine would weather very rapidly in a given environment, quartz which is also a primary mineral will persist under the same conditions within the same environment.

Some of the factors which tend to govern the relative stability of minerals to weathering are outlined as follows:

(a) *Climate of the Weathering Environment*

Climate is perhaps the most important factor influencing the rate of mineral weathering. For instance, the rate of weathering is generally faster in hot humid tropical climates than in temperate climatic environments. This is in the sense that under conditions of low rainfall, physical processes of weathering dominates, resulting in decreased particle-size with very little change in chemical composition of the minerals. The presence of more moisture such as occurs in humid tropical environments, encourages chemical as well as physical changes, resulting in new minerals and soluble products.

Weathering rates are generally more rapid in humid tropical areas, where the more resistant products of chemical weathering such as the hydrous oxides of iron and aluminium are prominent. Even quartz, the most resistant of the common macro-grained primary minerals, disappears in time under these conditions (Brady, 1974).

(b) *Particle-Size of Mineral Grains*

In general, a given mineral is more susceptible to decomposition when present in fine particles than in large grains. The much larger surface area of finely divided material presents greater opportunity for chemical attack. For example, quartz particles of sand-size are extremely resistant to chemical weathering. In contrast, clay-sized quartz, although not subject to ready decomposition, is not more resistant than many of the other minerals in similar-sized particles.

(c) *Chemical and Structural Characteristics of Minerals*

Chemical and crystalline characteristics determine to a large degree the ease of decomposition of minerals. Minerals such as gypsum (Ca $SO_4.2H_2O$) which are slightly soluble in water are quickly removed if there is adequate rainfall. Water charged with carbonic acid likewise dissolves less soluble minerals such as calcite and dolomite. Consequently, these minerals are seldom found in the surface horizons of humid environments.

Also dark-coloured primary minerals which often contain iron and magnesium in their structure and termed Ferromagnesian minerals, are more susceptible to chemical weathering than the feldspars and quartz. The presence of easily oxidizable Fe^{2+} and more easily hydrolysable cations such as Ca^{2+}, Mg^{2+} and K^+ in these minerals helps account for the more rapid break down of the ferromagnesian minerals like olivine, biotite, and pyroxenes.

Tightness or closeness of packing of the ions in the crystal units of minerals is also thought to influence their weathering rates. For example, olivine and biotite, which are relatively easily weathered have crystal units less tightly packed than those of zircon and muscovite which are structurally comparable minerals but are quite resistant to weathering.

(d) *Order of Crystallization of Minerals from Magma*

A "stability series" of the resistance of minerals to weathering proposed by Goldich (1938) coincides with the order of crystallization of minerals from molten magma also termed the Bowen series (Fig. 4.1). The series indicates that the first set of minerals to crystallize from molten magma are the most easily weathered, while the last set of minerals to crystallize are the most resistant to weathering. Apparently the least stable minerals are those that crystallize from a "melt" at the highest temperatures. This greater instability is related to their greater disequilibrium with the environment in the lithosphere – atmosphere interphase, the pedosphere (Buol et al., 1973).

Mineral Particles >2mm Diameter

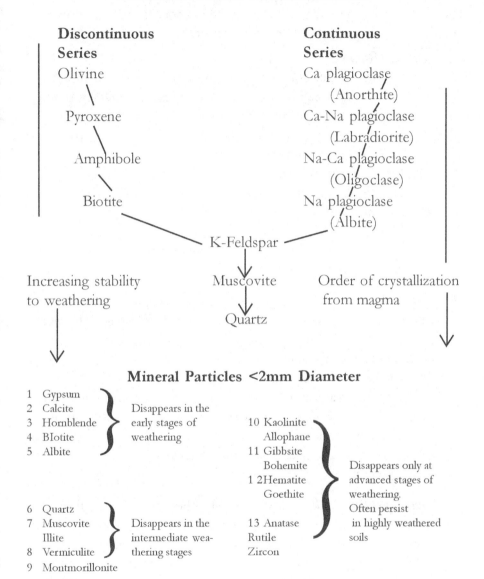

Fig. 4.1: *Stability to weathering of some common soil-forming minerals (After Goldich, 1938 and Jackson et al, 1984)*

Weathering Products of Common Soil Minerals in Rocks

The commonly known types of rocks each has specific types of minerals that they contain as in the following examples:

(1) **Igneous Rocks**
  (a) Granite usually contains quartz, potash feldspars (orthoclase/ microcline), sodium feldspars/plagioclase (oligoclase), biotite, amphiboles and iron ores (hematite, magnetite).
  (b) Basalt often contains clacic plagioclase (labradiorite), pyroxenes (mainly augite), olivine and iron ores.
  (c) Granodiorite contains quartz, K-feldspars, andesine, biotite, amphiboles and iron ores.
  (d) Diorite contains andesine, amphiboles, pyroxenes, biotite and little quartz and K-feldspars.
  (e) Gabbro contains labradorite, pyroxenes, olivine, amphiboles and little biotite and iron ores.

(2) **Sedimentary Rocks** – these vary a lot in composition due to the various associated cementing agents and sources of origin. However, some of the commonly occurring sedimentary rocks contain certain prominent minerals as follows:
  (a) Sandstones contain quartz (dominant), K-feldslpars and iron ores with clay minerals and carbonates as possible cementing agents.
  (b) Shales contain silt-size quartz, feldspars and muscovite with carbonates, mica and clay minerals as possible cementing agents.
  (c) Limestone contains mainly fine-grained calcite and some mica (glauconite).

(3) **Metamorphic Rocks** – these generally inherit the minerals from the original igneous or sedimentary rocks from which they are derived with some amount of modifications as in the following examples:
  (a) Gneisses contain quartz and feldspars as the predominant minerals in addition to oligoclase and biotite.
  (b) Schists contain predominantly micas (biotite, muscovite, sericite), pyroxenes, amphiboles, quartz and feldspars. Schistosis is due mainly to high mica content.
  (c) Amphibolites contain mainly hornblende (amphibole) and plagioclases (anorthite and oligoclase).

Judging from the foregoing list of commonly occurring soil forming rocks and the minerals they contain, the most common soil forming primary minerals include quartz, feldspars, micas and ferromagnesian minerals. When each of these minerals weather, the following products are produced under different conditions.

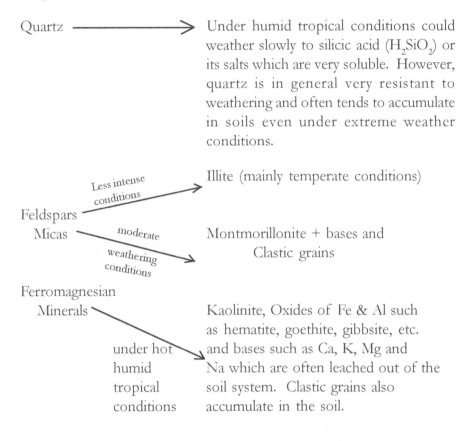

Quartz ──────────> Under humid tropical conditions could weather slowly to silicic acid (H_2SiO_2) or its salts which are very soluble. However, quartz is in general very resistant to weathering and often tends to accumulate in soils even under extreme weather conditions.

Illite (mainly temperate conditions)

Feldspars
Micas — Montmorillonite + bases and Clastic grains

Ferromagnesian Minerals

under hot humid tropical conditions — Kaolinite, Oxides of Fe & Al such as hematite, goethite, gibbsite, etc. and bases such as Ca, K, Mg and Na which are often leached out of the soil system. Clastic grains also accumulate in the soil.

Volcanic ──────────> Allophane ──> Halloysite ──> Kaolinite ──>
Ash, glass etc.   Oxides of Al & Fe.

Revision Questions

(1) Given granite and basalt as two parent rocks in a hot, humid tropical environment:

(a) Name 3 principal mineral species that could be present in each of the rocks.

(b) Which of the two rocks will weather faster and why?

(c) What would you expect the end products in the soil to be, and how would this affect the properties of the resulting soils.

(2)   Outline the basic groups of soil-forming minerals. List 12 minerals that are often most common in soil-forming rocks.

(3a) What is the relevance of weathering in pedology?

 (b) With the aid of suitable example, discuss Hydrolysis, Hydration and Oxidation-Reduction reactions as chemical processes of weathering.

References

Allen, B.L. & D.S. Fanning. 1983. *Composition and soil genesis.* In L.P. Wilding, N.E. Smeck & G. F. Hall (Eds.) Pedogenesis and Soil Taxonomy 1. Concepts and Interactions. Pp. 141-192.

Brady, N.C. 1974. *The nature and properties of soils.* 8th Edition, Macmillan Publishing Co., Inc., New York, U.S.A. 63pp.

Buol, S. W., F.D. Hole and R.J. McCracken, 1973. *Soil genesis and classification.* The Iowa State Univ. Press Ames, Iowa, 360pp.

Donahue, R.L., R.W. Miller and J.C. Shickluna. 1983. *Soils: an introduction to soils and plant growth.* Fifth Edition. Prentice-Hall Inc., Englewood Cliffs, New Jersey. 667pp.

Jackson, M.L., S.A. Tyler, A.L. Willis, G.A. Beurbean & R.P. Pennington. 1984. *Weathering sequence of clay-size minerals in soils and sediments.* I. Fundamental generalization. Journal of Physical and Colloid Chemistry 52, 1237-1260.

Goldich, S.S. 1938. *A study in rock-weathering. Journal of Geology 46:* 17-58.

5

Factors of Soil Formation

Introduction

Soil formation comprises two different processes. First, the changes from a consolidated mass (rock) not capable of supporting plant growth to the development of an unconsolidated (loose) layer of material that can support plants if climate is suitable and water is available; second, the changes occurring within the loose material as time passes. This latter process is also called soil development.

Five variables have been known to contribute to the processes of soil formation and soil development. Hans Jenny (1941, 1958), described the relationship between soil formation (S) and the "state factors" in an equation as follows:

$$S = f(cl, o, r, p, t, \ldots\ldots)$$

where

cl	=	climate of a region
o	=	organisms
r	=	relief or topography including hydrologic features such as water table
p	=	parent material
t	=	time
......	=	additional unspecified factors.

Pedologists have recognized the interdependence of the state factors of soil formation, but have often looked for situations in which, because all factors but one are "ineffective" in the landscape, the influence of the one variable factor is revealed. It is thus possible today to recognize sequences of soils which are dominated by single factors of soil formation.

These include, climosequences (climate dominant), biosequences (organisms), toposequences (topography/relief) lithosequences (parent material) and chronosequences (time). Thus, a lithosequence may be defined as a set of soils with property differences due solely to differences in parent material with all other soil-forming factors constant (Jenny, 1941).

We shall proceed to examine each of the state factors of soil formation in some measure of detail.

Parent Material

Parent material may be defined as the material from which soil is thought to have been derived. It is also defined as the unconsolidated, chemically weathered mineral or organic matter from which the A and B horizons (solum) of soils may have developed by pedogenic processes (Donahue *et al.*, 1983). Jenny (1941) also defined parent material as "the state of the soil system at time zero of soil formation" that is the physical body of soil and its associated chemical and mineralogical properties at the starting point of effect of a particular set of other soil forming factors. In other words, a previous soil or a previously weathered rock mass (saprolite) could be "parent material" by this definition and concept.

In general, parent materials influence soil formation by their different rates of weathering, the nutrients they contain for plant use and the dominant particle-size they contain. Also, the less developed a soil is, the greater will be the effect of parent material on the properties of the soil. Even the properties of well developed soils will be greatly influenced by the parent material, for instance, clay formation is favoured by a high percentage of decomposable dark (ferromagnesian) minerals and by less quartz content. Different soils will form from different types of parent material and result in **lithosequences** of soils. In the functional factorial analyses of Jenny (1941), a lithosequence is defined as a set of soils with property differences due solely to differences in parent material, with all other soil-forming factors constant. Expressed mathematically, the function is shown as:

$$S = f(pm) \; cl, o, r, t, ...$$

A classical example where lithosequences of soils occur in Nigeria is the Jos Plateau where at the same topography, climate, vegetation and time (Tertiary) obvious differences exist between soils derived from younger granites, older granites, newer basalts, older basalts, lateritized basalts, ryolite (Hill, 1978; Ojanuga & Awujoola 1981; Kparmwang, 1993).

Types of Parent Material

Three broad types of parent materials are generally recognized in soil formation. These include sedentary materials also called residual materials; unconsolidated sediments or transported materials and organic materials.

Sedentary or residual parent materials are those which are formed *in situ* from the weathering of bedrock (Plate 2). They are variously termed "In-situ weathered products", "weathered residuum", "weathered debris" or "saprolite", and they constitute the most common forms of parent material in Nigeria. Such materials could be of igneous, sedimentary or metamorphic origin.

Transported materials or unconsolidated sediments are materials which are either transported or deposited by water, wind or ice. Table 6.1 summarizes the agent, mode of deposition and the name of deposit of commonly transported or unconsolidated soil parent materials.

Table 5.1: *Agents, mode of deposition and name of deposit of transported parent materials*

Agent	Mode of Deposition	Name of Deposit	Group Name
Water	Stream (floodplain and Terraces)	Alluvium	
	Lake (deposited in)	Lacustrine	Fluvial Materials
	Lagoon	Lagoonal	or Deposits
	Ocean	Marine	Marine Deposit
Ice	Glacier	Till, Moraine	Glacial Deposits
		Outwash	or Materials
Wind	Wind	Loess (silty)	Aeolian
		Dunes (sandy)	Deposits
Volcanicity	Volcanic eruption	Volcanic Ash	Volcanic Deposits
Gravity	Gravity often aided by water and slope	Colluvium	Colluvial Deposits

Water deposited materials, also termed Fluvial Deposits are those transported by streams or rivers and are deposited on floodplains (levees and backswamps). They are termed *Alluvial materials* or alluvium (Fig. 6.1). Materials which are deposited in fresh lake water (non-saline) and which are later exposed either by lowering of the water level or by the elevation of land are termed *Lacustrine Deposits*. Extensive areas of lacustrine or lagoonal deposits occur in the Lake Chad Basin of Nigeria. Materials which are transported by salt or ocean water are termed *Marine Deposits*. On drainage, some of these deposits develop serious acidity problems and are sometimes called acid sulphate soils. Examples of soils derived from marine parent material are common in the mangrove swamp vegetation zone in Cross River, Akwa Ibom and Rivers States of Nigeria.

Materials which are deposited through the action of wind are termed Aeolian Materials. When such materials are mainly silty in content they are termed loess or Loessial Deposits. Extensive areas of loess deposits occur in the semi-arid region of northern Nigeria especially in Katsina (Funtua), Sokoto and Borno States. When the transported materials are mainly sandy, they are referred to as Dunes. Soils with dune parent material are commonly found in desert areas. In Nigeria, dune soils are common in Sokoto, Katsina and Borno States, which are at the periphery of the Sahara Desert.

Parent materials consisting of rock fragments and soil materials accumulated at the base of steep slopes as a result of gravitational action are termed Colluvial deposits or Colluvium (Fig.5.1). Colluviation is sometimes aided by the process of soil creep which is defined as the slow mass movement of soil and soil material down relatively steep slopes, primarily under the influence of gravity but facilitated by saturation with water and by alternate freezing and thawing.

Volcanic ash parent material occur in areas where there have been volcanic activities like in the Biu plateau area. The materials are composed of non-crystalline, tiny glass fragments, bits of easily weatherable feldspars and ferromagnesian minerals and varying amounts of quartz. Most of the volcanic ash deposits are andesitic (moderately basic in composition). Those which are highly siliceous are very vesicular and are commonly called *pumice*. Volcanic ash imparts rather distinctive properties on soils over a wide range of climatic conditions. One of the main features soils inherit from their volcanic ash parent materials is allophone, an amorphous aluminosilicate with which abundant organic matter is complexed in the upper solum.

Parent material which are deposited by moving ice or glaciers are variously called moraine, till and outwash. Moraine is an accumulation of glacial drift formed chiefly by the direct deposition from glacial ice; examples are ground, lateral, recessional and terminal moraines. Usually, the unstratified glacial drift deposited directly by the melting ice and consisting of clay, sand, gravel and boulders intermingled in any proportion is called till (Donahue *et al.*, 1983). These kinds of parent material are only found in temperate regions where glaciers abound.

The third major type of soil parent material is Organic Materials. These are accumulations of partially or wholly decomposed plant and animal debris which are characteristic of depressions in landscapes such as swamps, bogs and marshes. Anaerobic conditions promote the development of soils with organic parent material. Only little reserves of organic soils may occur in Nigeria within the mangrove swamp zones of southern Nigeria; particularly in Calabar, Port Harcourt and Lagos areas.

Climate

Climate is a dominant factor in soil formation mainly because of the effects of precipitation and temperature. Certain local variations in climate (microclimates) provide evidence of the influence of the climatic factor on soil formation within relatively small areas (Mosugu, 1989). However, the magnitude of climatic control on soil formation can best be seen by making comparisons that are global in scope. The influence of differences in climatic conditions result in the existence of climosequences of soils. In Nigeria, wide differences in soil pH, texture and nutrient contents occur among soils developed on similar parent material, of the same age and on similar topographic positions in the savanna (dry) and forest (wet) regions (Mosugu, 1989; Kparmwang & Esu, 1990; Moberg & Esu, 1991). Differences in soil properties are often due mainly to differences in climate, especially the rainfall regime.

Variables of Climate

The variables of climate which contribute to the development of climosequences of soils include:

(a) Precipitation (especially effective precipitation).

(b)   Temperature
(c)   Evapotranspiration
(d)   Wind velocity and direction
(e)   Sunshine percent
(f)   Seasonal variations of all of the above.

Precipitation followed by temperature are perhaps the most important variables of climate which influence soil formation. For instance, a soil is said to be "developed" when it has detectable layers (horizons), such as of accumulated clays, organic colloids, carbonates, or soluble salts that have been moved downward by water. The extent of colloid movement and the depth of deposition are determined partly by the amount and pattern of precipitation which produce the leaching action (Donahue et al., 1983). Soil acidity is also known to increase with increasing rainfall due mainly to the increased removal of cationic bases from the soil solum by leaching.

Weathering, leaching and erosion are more intense and of longer duration in warm, humid regions than in cold climates. In the hot, humid tropical region of southern Nigeria for instance, the soils are typically deep (often > 3m), high in clay content, well-drained, reddish in colour (due to the presence of oxidized iron) and are often low in essential plant nutrients because of increased rate of leaching.

The rate of organic matter decomposition is also almost directly controlled by temperature. Indeed, the influence of temperature on soil formation can be summarized by the Van't Hoff's temperature rule which states that "For every 10°C rise in temperature, the speed of a chemical reaction increases by a factor of two or three". Temperature is also the principal component in the calculation of potential evapotranspiration and thus has a large measure of control on the amount of effective rainfall.

Organisms

Plants, animals and other living organisms are active in soil formation. They aid in the breakdown of parent materials and in the formation and decomposition of organic matter in the soil. Vegetation affects soil formation by depositing organic materials in and on the soil. Differences in soils that have resulted primarily from differences in vegetation and the nature of other organisms present leads to the development of Biosequences of soils.

Role of Plants

In areas with grassland vegetation, the grasses typically have fibrous roots that die and decay each year and provide the surface soil with a good supply of organic matter. This imparts a dark colour to the soil surface. Grassland plants also tend to slow down the leaching of bases out of the surface soil by keeping those elements in the rooting zone, so that under grassland vegetation, the soil surface is usually only weakly acid or even alkaline in reaction. The relatively drier climates found in grassland regions also reduces leaching.

In the Nigerian savanna region however, where the vegetation is basically grassland, the organic matter content of the surface soils are relatively low, perhaps due to the rapid rate of mineralization of organic matter and the high degree of sheet erosion which washes away most of the organic residues on the soil surface. It is also known that in northern Nigeria, grasses are used for building of houses and grazing by animals (Esu & Ojanuga, 1985; Esu, 1986; 1987; Lombin & Esu, 1988).

Plant roots are not as concentrated near the surface in forests as in grasslands and so the surface soil under forest vegetation is usually lighter in colour than under grassland. Forests also produce large quantities of leaf and branch litter, which falls and often decay on the surface. Some of the products of this decay are organic acids, which enhance the solubilization and subsequent leaching of bases from the surface soil. This makes the soils to be relatively more acidic than grassland soils. Conifer litter generally produces more acid than does litter from deciduous trees.

Role of Animals

Soil animals can have very strong effects on soil formation. Burrowing animals such as moles, gophers, prairie dogs, rodents and particularly in Nigeria, earthworms, ants and termites are highly important in soil formation when they exist in large numbers. Soils that harbour many burrowing animals have fewer but deeper horizons because of the constant mixing within the profile usually termed Faunal Pedoturbation, which nullifies the organic colloids and clay movements downward.

The work of man in furrowing fields, removing hills, filling in low places, reducing (or increasing) nitrogen content of soil and accelerating erosion is familiar to us. Man as an animal, therefore contributes to soil formation by introducing new vegetation, erosion, soil amendments and modifying the climates of microenvironments through irrigation, drainage and the planting of shelter belts to check wind erosion.

Role of Microorganisms

Microorganisms contribute to soil formation by accelerating or slowing down the rate of decomposition of organic matter and forming weak acids that dissolve minerals faster than pure water. Some of the first plants to grow on weathering rocks are crust-like lichens which are a beneficial (symbiotic) combination of algae and fungi.

Collectively, micro-organisms contribute to soil development by contributing to weathering of rocks mostly through chemical processes; mineralization of organic matter to release plant nutrients and the production of humus which contribute to soil colour, soil structure development from microbial gums as well as nitrogen fixation by symbiotic and non-symbiotic bacteria.

Relief

Relief describes the lay of the land surface or the elevations or inequalities of the land surface considered collectively. It encompases and considers the geometrical relationships between mountains (hills), and connecting interfluvial slopes.

The influence of relief or topography is indirect as it serves mainly to redistribute matter and energy occurring from other soil-forming factors such as climate (moisture and temperature) and parent material. However, the indirect effect has a significant influence on soil formation. The altitude, shape of landscape, steepness and aspect of slope are important parameters to be considered.

Altitude

The effect of altitude (elevation) on soil development reflects itself through its influence on soil drainage which in turn materially influences soil development. Soils on high elevations are usually well drained whereas soils in lower topographic positions are usually poorly drained and of finer texture (Esu *et al.*, 1987).

Shape of Landscapes

Whether a landscape or landform is flat, undulating or hilly makes a difference in soil formation and development. Soils on hills or steep slopes receive little infiltration or effective precipitation due to accelerated run off. Soil profiles located on such hills or steep slopes are often very shallow, gravelly or stony due to minimal rate of weathering and removal of soil by erosion. These soils are often termed "skeletal" or "raw" mineral soils. Gently sloping landforms on the other hand allow ample infiltration of water into porous soils and develop into deep profiles (Fig. 5.1).

In flat landforms with flat or concave slopes, runoff is negligible or absent. Such slopes retain all the rain that fall on them and often receive additional amount of runoff water from adjacent uplands. Considerable amount of eroded materials may be deposited on these slopes.. Usually, poorly drained or hydromorphic soils and salt-affected or halomorphic soils typically occur in these landscape positions.

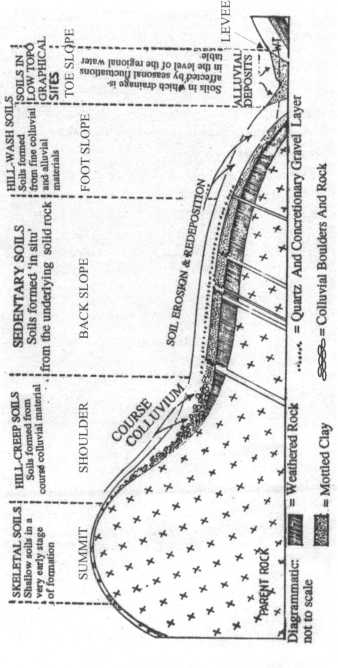

Fig. 5.1: Sedentary, colluvial and alluvial parent materials of soils along a typical toposequence in Western Nigeria. (Adapted from Symtb and Montgomery, 1962)

This is why it is often said that well developed or older soils are generally found on protected middle slopes while younger or weakly developed soils are found on lower slope positions.

An interfluve may have the relief positions described above; each relief position carrying different soil series which may have developed from the same parent material but differ in drainage. Such a group of soils is called a "Catena". However, it is becoming increasingly clear that slope may have been cut in different lithologic materials so that the soil series encountered on such slopes do not only vary in drainage but in parent material. Such a soil-slope sequence is called a "toposequence". The concept of toposequence is more applicable in Nigeria where soils occurring along the same slope vary in parent material as well as drainage (Symth & Montgomery, 1962; Esu, 1986; Esu *et al.*, 1987; 1991).

Aspect of Slope

Relief affects soil formation through its secondary influence of causing variations in exposure of land surfaces to the sun, wind and air-drainage. For instance, its been reported that in the northern hemisphere, south-facing slopes are warmer and drier than north-facing slopes. Also, because of the resulting high radiation received for photosynthesis, south-facing slopes tend to support a more luxuriant vegetal growth than north-facing slopes. The ensuing microclimate of south-facing slopes promotes rapid rate of organic matter decomposition, allow deeper percolation of water and hence favour the formation of deep soils.

Time

It is more realistic to think in terms of the length of time over which other factors (climate, organisms, parent material) have been influencing soil development than to think of time as an independent factor of soil formation. In other words, time factor is actually synthesized into other factors. The recognition of this fact probably prompted Crocker (1952) to formulate the following mathematical expression for soil formation:

$$S = \int_0^1 f(cl, o, r, p, \ldots)\, dt$$

Where S = soil formation, cl = climate
o = organisms, r = relief, p = parent
material t = time and ... = additional unspecified factors.

The expression states that soil formation is the integral function of all the factors of soil formation with respect to time. The beginning of soil formation can be taken to be the time zero (t = o) thereby progressing as factors or processes initiated by them operate through time.

Concept of "Time Zero"

Time like space may be regarded as continuous yet we may recognize a "time zero" for a given soil. Time zero is the point in time in which a pedologically catastrophic event is completed, initiating a new cycle of soil development. The catastrophe may be a sudden change in topography of a land surface or water table as caused by geologic uplift or tilting of a mass of the lithosphere. It could also be a rapid shift of a retreating slope due to geologic erosion or land forming.

Relative Stages of Soil Development

The influence of time on soil formation has long been recognized. Dokuchaev (1883) stated that the age of soil has begun from the moment the parent material emerged. Several scientists later began to study soil development on a time sequence. Hence a chronosequence, is a sequence of related soils that differ one from the other in certain properties primarily as a result of time as a soil-forming factor.

Today, the Davisian terms, Youth, Maturity and Old Age (senile), originally developed for description and identification of landscapes are also applied to soils. Many scientists discuss soil development as progressing through immature, mature and senile stages. The Immature stage is the beginning of soil formation with generally shallow soil profiles possibly with only AC horizons. Such soils are often low in clay content but high in the content of primary minerals. At the Mature stage of soil development, there is pronounced weathering of primary minerals to clay; constant organic matter content owing to equilibrium between gains and losses of soil materials and deep profile with ABC horizons. At the senile stages of soil development, the soils posses low weathering potential, declining organic matter content and declining fertility status.

Mohr and van Baren (1954), hypothesized five stages of soil development in the tropics, namely: Initial, Juvenile, Virile, Senile and Final. The Initial stage is the unweathered parent material; at the Juvenile stage, the weathering has commenced but much of the original mineral is still unweathered. At the Virile stage, there is pronounced decomposition of primary minerals tending to increased clay content and the soil has the maximum capacity to support vegetation or crops. At the Senile stage of soil development, the soils have low weathering potential with the soils being composed mostly of resistant minerals. At the Final stage, soil development has been completed and the soil has no weathering potential under the prevailing conditions.

Relative Rate of Soil Development

The development of the soil profile or horizonation requires time. Weathering, leaching, eluviation – illuviation etc., are time dependent. The time needed to form a unit depth of soil varies depending on other factors. For instance, for soil development on hard rock, the time may be measured in many centuries. In permeable, unconsolidated materials in a warm humid climate supporting forest vegetation, soil development may assume a rapid rate and time of development may be measured in 200 years to a few thousands of years. In permeable, unconsolidated material in a cold, dry climate, soil development would be slower than that for warm humid climate.

Revision Questions

1(a) What do you understand by the term Lithosequence of soils?
 (b) Discuss the role of unconsolidated sediments in soil formation.

2(a) List the variables of climate which contribute to soil formation.
 (b) Outline the role of micro-organisms in soil genesis.

3(a) Distinguish clearly between a catena and a toposequence of soils.
 (b) How do the shapes of landscapes influence soil formation?

4(a) State Crocker's equation of soil formation.
 (b) Outline the concept of "Time Zero" in the study of soil genesis.
 (c) Distinguish between Immature, Mature and Senile stages of soil development.

References

Crocker, R.L. 1952. *Soil genesis and the pedogenic factors*, Quart. Rev. Biol. 27: 139-168.

Dokuchaev, V.V. 1883. *Russian Chernozem* (Russkii Chernozem). Translated from Sci. Trans. Jerusalem, 1967. Available from U.S. Dept. Commerce, Springfield, Va.

Donathue, R. L., R. W. Miller & J.C. Shickluna. 1983. *Soils: an introduction to soils and plant growth*. Fifth Edition. Prentice-Hall Inc., Englewood Cliffs, New Jersey. 667pp.

Esu, I. E. 1986. Morphology and classification of soils of the Nupe sandstone formation in Niger State, Nigeria. *Samaru Journal of Agricultural Research*, 4 (1 & 2): 13-23.

Esu, I. E. 1987. Fertility status and management of some upland basement complex soils in the Nigeria savanna region. *Nigerian Journal of Soil Science*, 7: 155-183.

Esu, I. E, I. J. Ibanga & G. Lombin. 1987. Soil-landscape relationships in the Keffi plains of northern Nigeria. *Samaru Journal of Agricultural Research*, 5(1 & 2): 109-123.

Esu, I. E., A. C. Odunze & J. P. Moberg. 1991. Morphological, physico-chemical and mineralogical properties of soils in the Talata – Mafara area of Sokoto State. *Samaru Journal of Agricultural Research*, 8: L 41-56.

Hill, I. D. (ed) 1978. *Land resources of central Nigeria: Agricultural Development Possibilities. Vol. 2B. The Jos Plateau*. LRD. Tolworth Tower, Surbiton, Surrey, England. Pp.9-11 and 14-16.

Jenny, H. 1941. Factors of soil formation. McGraw-Hill, New York.

Jenny, H. 1958. Role of the plant factor in the pedogenic functions. *Ecology* 39:5-16.

Kparmwang, T. 1993. *Characterization and classification of basaltic soils in the northern guinea savanna zone of Nigeria*. Unpublished Ph.D Thesis. Dept. of Soil Science, ABU, Zaria. 176pp.

Kparmwang, T. & I. E. Esu. 1990. Mineralogical studies of alfisols containing plinthite in two ecosystems in Nigeria. *Savanna*, II(1): 109-121.

Lombin, G. and I. E. Esu. 1987. Characteristics and management problems of vertisols in the Nigerian savanna.In:Jutzi, S.C.*et al* (eds). *Management of Vertisols in Sub-Saharan Africa,* ILCA, Addis Ababa, Ethiopia. Pp.293-307.

Moberg, J. P. & I. E. Esu. 1991. Characteristics and composition of Nigerian savanna soils. *Geoderma,* 48: 113-129.

Mohr, E. C. J. & F.A. Van Barren. 1954. *Tropical soils: A critical study of soil genesis as related to climate, rock and vegetation.* The Royal Tropical Institute, Amsterdam. Interscience, New York. 481pp.

Mosugu, M. 1989. *A bio-climosequence study of upland soils developed on older granites in Nigeria.* Unpublished M.Sc Thesis. Dept. of Soil Science, ABU, Zaria 132pp.

Ojanuga, A. G. & A. I. Awujoola. 1981. Characteristics and classification of the soils of the Jos Plateau, Nigeria. *Nigeria Journal of Soil Science,* 2:101-119.

Symth, A.J. & R.F. Montgomery. 1962. Soils and land use in central Western Nigeria. Govt. Printer, Ibadan, Western Nigeria. 265pp.

6

Soil Forming Processes

Introduction

The process of soil formation is best visualized as a complex sequence of events which include both complicated reactions and comparatively simple rearrangements of matter that intimately affects the soil in which it operates (Buol *et al.*, 1973). Numerous events may take place simultaneously or in sequence to mutually reinforce or contradict each other (Rode, 1962; Simonson, 1959).

The universally acknowledged external factors which influence soil formation have earlier been discussed (chapter 5) and include climate, parent material, living organisms, topography and time. The internal processes which are often influenced by the external factors can be classified into four categories as follows:

(a) Additions or gains of organic and mineral materials to the soil body.
(b) Losses of these materials from one point to another within a soil body.
(c) Translocations of materials from one point to another within a soil body.
(d) Transformation of mineral and organic substances within a soil body.

Some of the soil forming processes, also termed pedogenic processes are discussed below and an attempt at their categorization is summarized in Table 6.1. It is however, pertinent to add that the pedogenic processes are not easily compartmentalized as stated above. This is in the sense that addition and losses of materials within a soil body may occur together in a single complete process; addition and translocation may occur together and so on. In the discussions which follow, attempt will therefore be made to treat individual and some important pedogenic processes without necessarily grouping them into additions, losses, translocation and transformation processes. The emphasis here will be on the mode of operation of the individual or group of processes in soils:

Eluviation – Illuviation

The terms eluviation and illuviation are analogous to the words emigration and immigration respectively (Buol *et al.*, 1973). The processes involve the movement of material, usually clay, organic material or Fe & Al oxyhdroxides in suspension from one part of a soil profile (usually an upper, say E horizon) and deposition in another (usually a lower B or Chorizon) within a given soil profile.

Two aspects of eluviation are mobilization and translocation, while illuviation also involves translocation processes and their interruption by immobilization of the moving materials. The process is impeded in an environment where Al^{3+} and Ca^{2+} saturate the soil, as they tend to cause flocculation. Under tropical conditions, therefore, eluviation-illuviation processes are enhanced in moderately acid environments dominated by the cations K^+, H^+ and Na^+. The process also requires the occurrence of a network of pores and fissures through which the materials can move down in suspension. The occurrence of pores and fissures implies that a wetting and drying cycle must exist within the environment of the process; the pores remaining open during the dry cycle and particles in suspension moving down the pores during the wet cycle.

Eluviation-Illuviation process when it involves clay, often lead to the development of argillic (Bt) horizons and the processes are variously termed Lessivation, Deargillation – Argillation or Argeluviation – Argilluviation processes.

Table 6.1:   *Some processes of soil formation that are complexes of subprocesses and reactions (Adapted from Buol et al., 1973)*

Term	Fourfold Categori- zation	Brief Definition
1a. Eluviation	3	Movement of material usually fine solids in suspension out of a portion of a soil profile as in an albic horizon (E-horizon)
1b. Illuviation	3	Movement of material as above into a lower portion of soil profile as in an argillic or spodic horizon (B horizon)
2a. Leaching (depletion)	2	General term for washing out soluble materials in solution from the solum.
2b. Enrichment	1	General term for addition of material to a soil body.
3a. Erosion, surficial	2	Removal of material from the surface layer of a soil.
3b. Cumulization	1	Aeolian and hydrologic additions of mineral particles to the surface of a soil solum.
4a. Decalcification	3	Reactions that remove calcium carbonate from one or more soil horizons.

Term	Fourfold Categori- zation	Brief Definition
4b. Calcification	3	Processes including accumulation of calcium carbonate in Ck and possibly other horizons of a soil.
5a. Salinization	3	The accumulation of soluble salts such as sulphates and chlorides of calcium, magnesium, sodium, and potassium in salty (salic) horizons.

5b. Desalinization	3	The removal of soluble salts from salic soil horizons.
6a. Alkalization (solonization)	3	The accumulation of sodium ions on the exchange sites in a soil.
6b. Dealkalization	3	The leaching of sodium ions and salts (desolodization) from natric horizons.
7a. Lessivage	3	The mechanical migration of small mineral particles from the A to the B horizons of a soil, producing in B horizons relative enrichment in clay (*argillic horizons*).
7b. Pedoturbation	3	Biologic, physical (freeze-thaw and wet-dry cycles) churning and cycling of soil materials, thereby homogenizing the solum in varying degrees.
8a. Podzolization (silication)	3,4	The chemical migration of aluminium and iron and/or organic matter, resulting in the concentration of silica (i.e., silication) in the layer eluviated.
8b. Laterization (desilication, ferralization, ferritization, allitization)	3,4	The chemical migration of silica out of the soil solum and thus the concentration of sesquioxides in the solum (goethite, gibbsite, etc.), with or without formation of ironstone (laterite; hardened plinthite) and concretions.
9a. Decomposition	4	The breakdown of mineral and organic materials.
9b. Synthesis	4	The formation of new particles of mineral and organic species.

10a. Melanization	1,3	The darkening of light-coloured initially unconsolidated mineral materials by admixture of organic matter (as in a dark A or mollic or umbric horizon).
10b. Leucinization	3	The paling of soil horizons by disappearance of dark organic materials either through transformation to light-coloured ones or through removal from the horizons.
11a. Littering	1	The accumulation on the mineral soil surface of organic litter and associated humus to a depth of less than 30cm.
11b. Humification	4	The transformation of raw organic materials into humus.
11c. Paludization	4	Processes regarded by some workers as geogenic rather than pedogenic, including the accumulation of deep (>30cm) deposits to organic matter as in mucks and peats (Histosols).
11d. Ripening	4	Chemical, biological, and physical changes in organic soil after air penetrates the organic deposit, making it possible for microbial activity to flourish.
11e. Mineralization	4	The release of oxide solids through decomposition of organic matter.
12a. Braunification, Rubifaction Ferrugination	3,4	Release of iron from primary minerals and the dispersion of particles of iron oxide in increasing amounts; their progressive oxidation or hydration, giving the soil mass brownish, reddish brown, and red colors respectively.

| 12b. Gleization | 3,4 | The reduction of iron under anaerobic "waterlogged" soil conditions, with the production of bluish to greenish grey matrix colours, with or without yellowish brown, brown, and black mottles, and ferric and manganiferous concretions. |

Note on Fourfold categorization: 1=Addition; 2=Losses; 3=Translocation and 4=Transformation processes.

Leaching and Enrichment

The processes of leaching and enrichment are analogous to eluviation and illuviation. The major difference between the two sets of translocation processes is that the process of leaching involves movement in complete aqueous solution downward in true solution and not in suspension like in the eluviation process. In leaching, the losses involve mainly plant nutrient ions, which accumulate in other sections of the landscape in the process of enrichment. Cations often leached under humid tropical conditions include Ca^{2+}, Na^+, K^+, and Si^{4+}, while the anions often leached include Cl^-, NO_3^-, SO_4^{2-}, $H_2PO_4^-$ and HCO_3^-.

Decalcification and Calcification

Under less arid conditions where in a whole year, there exists a moisture surplus, the more mobile soluble constituents (Na^+,K^+) move down through the profile rather than to the surface. On the other hand, the less soluble calcium compounds are mobilized but reprecipitate lower in the soil, possibly as a Bk or Ck horizon. Because of the strong biological activity in these soils, considerable carbon dioxide is liberated by respiration. This enables calcium carbonate to be converted into the soluble bicarbonate which can move down through the profile in a Decalcification process as follows:

$$Ca\,CO_3 + H_2O + CO_2 \rightleftharpoons Ca(HCO_3)_2.$$

In areas of low rainfall and grassland cover, reprecipitation is encouraged by a decline with depth in biological activity and therefore lower CO_2 content, and by evaporation which affects much of the profile during the prolonged periods of dry weather. The above Decalcification reaction is then reversed with the formation of secondary deposits of crystalline calcium carbonate (calcite) sometimes as infilled animal burrows or filaments:

Leaching (Decalcification)

$$Ca^{2+} (aq) + 2 HCO_3^- (aq) \rightleftharpoons CaCO_3(s) + CO_2(g) + H_2O$$
Enrichment (Calcification)

Along a strongly sloping landscape, the process will occur as in Fig. 6.1:

Formation of
CaCO$_3$ at break of slope

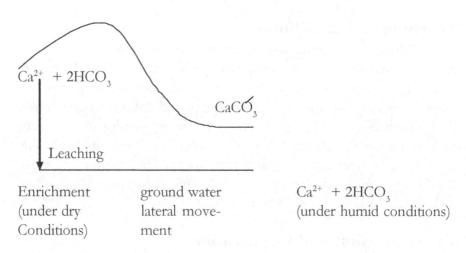

Fig. 6.1: *Decalcification and Calcification Processes along a Hill slope*

Dealkalization and Alkalization

Dealkalization refers to the removal of Na$^+$ ions from the exchange sites. This process also involves much dispersion of clay. Dispersion occurs when the Na$^+$ ion becomes hydrated. Much of the dispersion can be eliminated if Ca^{++} and/or Mg^{++} ions are concentrated in the water used to leach the alkali soil as they can replace the Na$^+$ on the exchange complex.

Alkalization on the other hand, involves the accumulation of sodium ions on the exchange sites of clay. According to Buol *et al.*, (1973), all the cations in solution engage in a reversible reaction with the exchange sites on the clay and organic matter particles. This reaction can be represented by the following equation, where X represents clay or organic matter exchange sites as follows:

$$\text{CaMg 2Na X} \rightleftharpoons \text{Ca}^{++} + \text{Mg}^{++} + 2\text{Na}^+ + \text{X}^{-6} + 3\text{CO}_3^- \rightleftharpoons \text{Na}_2\text{CO}_3 + \text{MgCO}_3 + \text{CaCO}_3$$

Since Na_2CO_3 is more than 100 times as soluble as Ca or Mg carbonate, it is expected that much of the Ca^{++} and Mg may be precipitated from solution before Na is precipitated. Thus, the concentration of Na^+ left in solution to react with the exchange sites is high. The precipitation of the Ca^{++} and Mg^{++} in carbonates takes place first as the soil dries out. The Na^+ then attaches itself to the clay and organic matter exchange sites and the soil becomes sodified or solonized. Hence, the processes of Alkalization and Dealkalization are also termed Sodification, Solodization or Solonization and Desodification, Desolodization or Desolonization respectively.

Allitization and Deallitization

This process involves the release and/or disappearance of $Al(OH)_3$. Allitization occurs under intense weathering conditions in hot, humid climates where the annual precipitation exceeds 600-700mm and the bulk of it happens to fall during one or two peaks during the wet season. Note that with lower annual precipitation and its even distribution through the seasons, the **siallitization** process of mineral mass transformation rather dominates. Hence allitization is directly related to heat and moisture especially if the soil parent material is basic e.g. basalt.

$$\text{Ca Al}_2 \text{Si}_2 \text{O}_8 + 2\text{H}^+ + 6\text{H}_2\text{O} \longrightarrow 2\text{Al(OH)}_{3(s)} + 2\text{Si(OH)}_4 + \text{Ca}^{2+}$$
(Anorthite)   (Gibbsite)   leached   leached

In general, Deallitization occurs if the $Si(OH)_4$ is not easily removed or leached from the soil, and Kaolinite will then easily form as in a granitic rock environment. If we have high levels of $Si(OH)_4$ in the soil environment, the synthesis of aluminosilicate gels (allophone or imogolite) may result as follows:

$$\text{n Al (OH)}_3 + \text{n Si(OH)}_4 \longrightarrow \text{Al}_2\text{Si}_2 \text{(OH)}_6 \text{O}_4 3\text{H}_2\text{O}$$
(Allophane)

or

$$\text{Al}_3\text{Si}_2 \text{(OH)}_8 \text{O}_6 \text{ (no water)}$$
(Imogolite)

Allophane may on loss of associated water molecules become halloysite which could further loose its water of hydration to become Kaolinite.

Salinization and Desalinization

Salinization operates chiefly in subhumid, semiarid and arid regions and some coastal humid regions, wherever depressions are enriched in salts faster than they are leached. Salt accumulation is preferential in depressional soils high in content of clay and low in permeability with reduced leaching. The predominant salts involved in soil salinization are sulphates and chlorides, but nitrates and borates occur rarely.

Desalinization is most frequently used with reference to the removal by leaching of soluble salts from horizons or total soil profiles that have previously contained enough soluble salt so that plant growth was impaired. Therefore, it is a process that can be active only after soluble salts have accumulated, that is, salinization.

Podzolisation

Podzolisation may be defined as the process by which sesquioxides are translocated in a soil profile and can be explained in part by the solubilities of ferrous (Fe^{++}) and ferric (Fe^{+++}) iron (Stobbe and Wright, 1958). While the soluble Fe^{++} forms are at the sites of eluviation, the insoluble Fe^{+++} forms are at the point of illuviation.

It is actually a pedogenic process which is common in the temperate region under coniferous (boreal) forest and involves a bundle of processes which include; accumulation of organic matter, leaching and acidification; weathering; translocation of Fe and Al (with some P, Mn and clay) from the O, A and E to B horizons; immobilization of humic and fulvic acids (and some clay) in the B horizon; pelleting of humus coatings; reduction in bulk density and cementation.

Eluviation Illuviation in the podozolisation process is mainly through metal-organo complexes (movement of Fe and Al oxides with organic matter).

With this chelation, the Fe-oxyhydroxide coatings on quartz grains which initially are insoluble then become soluble and are carried down the profile. According to Franzmeier & Whiteside (1963), the amount of Fe and Al accumulated in a Spodosol through the process of podzolisation, may be far more than has been biocycled in the history of the soil. The source of the bulk of these two elements in the spodic horizon, therefore, has come from weathering of ferromagnesian minerals, feldspars, illite and chlorite in the A and E horizons, where pH usually drops below 5.0. Clay lessivation is sometimes viewed as a precursor of podzolisation. After clay is eluviated from surficial horizons, albic and spodic horizons now form in the coarser and finer residues respectively.

Podzolisation in the tropics and subtropics is limited and is mainly confined to highland areas. There, the conditions for litter accumulation are more favourable, likewise for that of acid organic substances destroying the mineral part of the soil. In addition, it may be suggested that in such mountainous regions, podzolisation is related to the siallitic stage of rock weathering.

Lateritization

The process of lateritization also termed ferralization, latosolization or plinthization, is mostly vividly manifested only under tropical humid conditions, where some forms of iron transform into recent geological rock. During the process, neoformations of iron, chiefly as ferruginous quartz concretions are formed. These concretions then accumulate at different depths from the soil surface and form a layer of concretions which are later cemented by new supplies of iron into continuous layers of different thickness often termed plinthite when still soft or petroplinthite, ironstone or laterites when they have hardened irreversibly on exposure to wetting and drying cycles.

For lateritization to occur, the following conditions must be met:
(a) an additional inflow of ferruginous water (lateral or vertical);
(b) the change in the reaction of the medium along the path of the

water (from acid to alkaline) or sharp changes between oxidation and reduction;

(c)   frequently, it requires that the coarse-sand or gravel texture varies along the profile (from light to heavy);

thus, increasing the lateral migration of water, while its vertical migration decreases. These requirements implies that seasonal or constant excessive moistening is vital for lateritization to occur. This is why laterite layers can also form in local depressions where there is an iron-rich surface and additional top water inflow.

In general, lateritization also refers to processes that remove silica (desilication) from the soil and thus results in the concentration of sesquioxides in the solum (goethite, gibbsite, etc.) with or without formation of ironstone (laterite, plinthite) and concretions. High temperatures and extreme regions, favour rapid desilication and accumulation of iron (ferritization) immobilized in ferric oxide forms under oxidizing conditions as in the following example:

$$KAISi_3O_8 \text{ weathered} \longrightarrow K^+ + Al(OH)_3 + 3Si(OH)_4$$

$$\text{leached} \quad \text{Gibbsite} \quad \text{aqueous}$$

$$\text{(residual)}$$

Also, under conditions of high leaching, acidification, moderate drainage and semi-arid or subhumid conditions we can have the reaction

$$4KAISi_3O_8 + 4H^+ + H_2O \longrightarrow Al_4Si_4(OH)O_{10} + 8Si(OH) + 4K^+$$

$$\text{Kaolinite}$$

Gibbsite formation is best favoured at the break of slopes where much of SiOH is often removed especially under acidic (pH < 5.5) conditions.

Gleization

The process of gleization occurs as a result of over-moistening of soil mass by groundwater when the backing water fills all capillaries. Periodical surface over moistening (in rainy seasons for example) helps water fill in some noncapillary passages. This process is especially vigorous when the table of groundwater is close to the surface or the water is stagnant. When the groundwater is deeper, gleying diminishes and only active in places where the soil mass's capillaries are moistened most heavily.

Gleying, therefore, usually develops with insufficient supplies of oxygen when there are anaerobic conditions in the soil mass. The bacterial flora changes and species develop which breathe oxygen from mineral and organic compounds. With no supply of oxygen, some oxides, especially iron oxides, undergo reduction (e.g. $Fe_2O_3 \longrightarrow FeO$). These changes are outwardly expressed in form of specific changes in soil colour such as grey, blackish, greenish or bluish matrix colours. Mottling within the zone of fluctuating water table position as well as yellow or red streaks within root channels are also observed.

Generally, two forms of gleys are recognized. These are:

(a) Pseudogleys, which are gleys occurring only within root channels and pores, in which there is rapid water movement within the main soil matrix but the pores are gleyed due to the reduction of Fe^{3+} to Fe^{2+} which takes place only within the root channels and pores which are rather slowly drained.

(b) Stagnogleys, which are due to stagnant water table position. In this case, oxidizing conditions occur in the pores, and root channels, while the soil matrix is reduced.

In general, we expect to find the following conditions in a true gley soil:

(i) High water table and therefore low degree of leaching, except perhaps lateral leaching.

(ii) The process of enrichment is therefore the common pedogenic phenomenon.

(iii) No erosion but accumulation because they often occur in level, depressional areas. The enrichment will be in bases and silica and include Mg, Ca, K, Na, Fe^{2+}, $Si(OH)_4$. There would also be accumulation of clay and organic matter.

(iv) Smectite formation or Smectification (Montmorillonitization) by the process of Neoformation from Kaolinite when equilibrium is established could be a dominant process as shown in the following equation:

$$3\frac{1}{2}Al_4Si_4(OH)_8O_{10} + 2Mg^{2+} + 2Fe^{2+} + 3Ca^{2+} + Si(OH)_4$$
Kaolinite

$$2M^+(Al\ Mg\ Fe)_4\ (Si_{7.33}\ Al_{0.66})\ (OH)_4\ O_2 + 8H^+ + 26H_2O$$
Smectite

The bases and silica are leached from the uplands into the poorly drained depressions. The process could also lead to Dekaolinization process as seen in the equation above.

(v)   Low amounts of Fe - oxyhdroxides; because Fe is not stable under conditions of poor drainage.

(vi) Primary minerals notably micas (biotite, muscovite) will be much more stable under gley conditions than in well drained soils because there is no tendency for the breakdown of their interlayer since there are abundance of cationic bases like K^+, Ca^{++}, Mg^{++} and Na^+ within the environment already.

Ferrolysis

The term Ferrolysis is derived from *ferro*(us) and *lysis* and was first coined by Brinkman (1970) as a short term for disintegration and solution in water by a process based upon the alternate reduction and oxidation of iron. It is a hydromorphic soil forming process involving seasonally wet pseudogley and gley soils.

The process involve reactions in which a soil's cation exchange capacity is destroyed due to exchange reactions involving iron in seasonally alternating cycles of reduction and oxidation. The essential reactions which take place may be summarized as follows:

Under   Flooded conditions   (Wet season)

$$2\ Fe\ OOH + 4e^- \longrightarrow 2Fe^{2+} + H_2O + 1\frac{1}{2}O_2$$
Flooding

Under these conditions, the pH of the soil environment increases, reducing soil acidity due to the following reactions which increases the bases in soil

$$Fe^{2+} + \begin{array}{|c|} Ca^{2+} \\ Mg^{2+} \\ K^+ \\ Na^+ \end{array} \longrightarrow \begin{array}{|c|} Fe^{2+} \\ Ca^{2+} \\ Fe^{2+} \\ Fe^{2+} \end{array} + Mg^{2+} + K^+ + Na^+ \downarrow$$

Clay mineral   clay mineral

When the soil dries out (Dry season), the following process occurs:

$$\begin{array}{|c|} Fe^{2+} \\ Fe^{2+} + H^+ \\ Ca^{2+} \end{array} \xrightarrow{} \underset{goethite}{Fe(OH)_3} + \begin{array}{|c|} Ca^{2+} +Al^{3+} \\ H^+ \\ H^+ \\ H^+ \\ H^+ \end{array} \begin{array}{l} \text{(Release from} \\ \text{the octahedral} \\ \text{lattice edges of} \\ \text{clay mineral)} \end{array}$$

Under these conditions the ferrolysed horizon becomes quite acidic in reaction. The clay mineral on which the H^+ are Adsorbed become quite unstable and will release Al^{3+} from the octahedral lattice edges further making the soil more acid.

When Flooded Again, the reactions are:

$$2Fe^{2+} + \begin{array}{|c|} Ca \\ Al^{3+} \\ H \end{array} \xrightarrow{} \begin{array}{|c|} Fe^{2+} \\ Al^{3+} \\ H^+ \end{array} + Ca^{2+} + Fe^{2+}$$

At this stage, the soil begins to lose Fe^{2+} and the colour of the soil begins to change, albic/spodic horizon develops and the soil becomes more acidic as the conditions continue year after year. This happens a lot in terrace soils which are seasonally flooded and liming therefore becomes a must if the soils are to be cultivated.

It must be noted that the process of ferrolysis is distinct from podsolization, argilluviation or gleization. Podsolization is not dependent upon seasonal reduction; argilluviation (Dudal, 1968) does not involve clay destruction; gleization does not require elimination of reaction products or clay destruction and may in some cases involve the reverse. However, the concept of ferrolysis would account for such diverse unresolved questions as the anomalously low cation exchange capacity of many topsoils under long-continued seasonal rice cultivation; the potassium fixing nature of some pseudogleys and the presence in them of Al-interlayered clays (soil chlorites); and the apparently greater speed of podsolization in seasonally wet conditions.

Where ferrolysis is active, the ferrolysed horizon is near neutral in reaction in the reduced condition but acid when oxidized (Esu; 1989). There is an appreciable content of exchangeable ferrous iron in the reduced phase and of exchangeable aluminium when the soil is oxidized.

Pedoturbation

Pedoturbation is a term used as a synonym for "soil mixing" and represents a local cyclic movement of soil materials either by animals, plants, or a number of phenomena such as earthquakes, movement of gas, expansion and contraction of silicate clays and freezing and thawing of permafrost soils among others (Hole, 1961; Buol et al.1973; Esu, 1995).

Specific terms are recognized to describe pedoturbation activities of animals, plants and various phenomena. For instance, *Faunalpedoturbation* refers to the mixing of soil by animals such as ants, earthworms, rodents and man, *Floralpedoturbation* refers to soil mixing by plants, as may occur in tree fall by wind throw. *Congelipedoturbation* is a term which describes the mixing of soil materials by freezing and thawing. *Argillipedoturbation* is the mixing of soil by expansion and contraction of smectite-clay rich soils, as in the self-mulching vertisols of northeastern Nigeria (Lombin & Esu, 1988). *Gravipedoturbation* is the mixing of soil materials by non-catastrophic materials within a soil creep. *Aeropedoturbation* is the mixing of soil materials within a soil profile by the movement of gas, as in the case of moving air bubbles. *Aquapedoturbation* is soil mixing by water within the soil profile such as by upwelling seepage waters in a soil profile. *Crystalpedoturbation* refers to soil mixing by the growth and wasting of crystals in the soil. *Seismipedoturbation* is the mixing of soil materials within a soil profile by earthquakes.

Revision Questions

1.   Write extensively on the following sets of soil forming processes and point out in each case how they may lead to the development of diagnostic subsurface horizons in soils:
      (a)   Eluviation – Illuviation
      (b)   Podzolisation
      (c)   Leaching – Enrichment.

2.   Describe by means of chemical equations how smectification may take place by neoformation from kaolinite.

3. Describe in chemical terms the processes covered by the term "ferrolysis".
   (a) Under which conditions will such processes take place?
   (b) How will they affect the edaphological properties of the soil in which they occur
   (c) How can human beings accelerate or decelerate this development?

References

Brinkman, R. 1970. Ferrolysis, a hydromorphic soil forming process. *Geoderma*, 3: 199-206.

Buol, S. W., F.D. Hole & R. J. McCracken, 1973. *Soil genesis and classification*. The Iowa State Univ. Press, Ames, Iowa, 360pp.

Dudal, R. 1968. Definitions of soil units for the soil map of the world. Soil Map of the World, FAO/UNESCOI project. *World Soil Resources Report*, 33.

Esu, I. E. 1989. A pedological characterization of soils of the Hadejia alluvial complex in the semi-arid region of Nigeria. *Pedologie*, XXXIX − 2, 171-190.

Esu, I. E. 1995. Effect of earthworm *(Hyperiodrilus africanus)* pedoturbation on selected soil properties in the levees of the cross river, south eastern Nigeria. *African Soils* (Special Edition). Vol. XXVIII: 363-374.

Franzmeier, D.P. & E.P. Whiteside, 1963. A chronosequence of podzols in northern Michigan. I. Ecology and description of pedons. Mich. State Agric. *Exp. Sta. Quart. Bull 46:2-20*.

Hole, F. D. 1961. A classification of pedoturbations and some other processes and factors of soil formation in relation to isotropism and anisotropism. *Soil Sci.* 91: 375-377.

Lombin, G., & I. E. Esu. 1988. Characteristics and management problems of vertisols in the Nigerian savanna. In: Jutzi, S. C. *et al.*, (eds). *Management of Vertisols in Sub-Saharan Africa*. ILCA, Addis Ababa, Ethiopia, pp. 293-307.

Rode, A. A. 1962. *The process of podzolization. Soil Science* (Pochvovediniye). (Transl. from Russian by A. Gourevich). Israel Prog. For Sci. Trans.,

Jerusalem.   Available U.S. Dept. Commerce, Washington D.C.

Simmonson, R. W. 1959.   Outline of a generalized theory of soil genesis. *Soil Science Soc. Am Proc.* 23: 152-156.

Stobbe, P. C. & J. R. Wright.   1959.   Modern Concept of the genesis of Podzols.   *Soil Sci. Soc. Am. Proc. 23: 161-164.*

7
Soil Morphological Properties

Introduction

Pedologists are trained to differentiate between various soil bodies within any given landscape using such broad criteria as:

(a) landforms (floodplains, terraces, residual uplands, dunes etc.);

(b) surface drainage patterns and eroded areas;

(c) erosional forms and eroded areas;

(d) landuse patterns and landuse boundaries;

(e) major types of natural vegetation;

(f) rock outcrops, etc.

However, detailed information on the morphological characteristics of a representative soil body or soil type is best evaluated from the *in situ* examination of a soil profile . Such a profile to be examined and described must according to Soil Survey Staff (2006) meet the following criteria:

(i) A freshly dug profile pit as shown in **Plate 5**, with dimensions of 2m x 1.5m x 2m is desirable, unless the depth to an impenetrable layer or ground water table is shallower than 2m.

(ii) Pits should be cited at least 50m from road cuts, quarries, houses and other features likely to disturb or contaminate the soil profile.

(iii) As much as possible, two sides of the pit from which observations will be made should be kept free of spoil and untrampled.

(iv) Normally, at least two pits in each major kind of soil body should be examined.

Plate 5: *A freshly dug profile pit with marked horizons for description*

In describing each profile pit, first a general site information regarding the characteristics of the site and a detailed characterization of horizon properties must be recorded.

General Site Information

The site information which should be made and recorded should include the following:
(a) Sampling data and period
(b) Geographical location of profile pit
(c) Soil mapping unit identification
(d) Profile pit identification No.
(e) Elevation (if known) or measure with a GPS
(f) Climatic information of the area
(g) Soil parent material
(h) Geology and Geomorphology
(i) Topographic position and slope
(j) Vegetation
(k) Erosion hazard (type and degree)
(l) Drainage conditions
(m) Surface soil features e.g. stony, hummocky, etc.

(n)   Depth to water table position

(o)   Depth to any impenetrable layer.

Details of the style or method of describing the above listed parameters can be found either in the Soil Survey Manual (Soil Survey Staff, 1993) or Guidelines for Soil Profile Description (FAO, 1977) or the Field Book for Describing and Sampling Soils (Schoeneberger et al., 2002).

Description of Individual Soil Horizons

For each profile pit, the various horizons or soil layers should be identified and for each horizon or layer, the following soil morphological characteristics should be noted:

(a)   Master horizon designations and subhorizon symbols.

(b)   Depth of top and bottom of each horizon; this should be continuous e.g. 0-15cm, 15-40cm, 40-85cm, 85-110cm, 110-200cm.

(c)   Soil colour (state whether in moist or dry conditions).

(d)   Abundance and colour of mottle(s), if any.

(e)   Soil texture

(f)   Soil Structure

(g)   Soil consistence (wet, moist or dry).

(h)   Cutans

(i)   Pores

(j)   Roots

(k)   Included materials e.g. concretions, nodules, stones, minerals, animal activity, etc.

(l)   pH (field test kit)

(m)   Horizon boundary characteristics.

Horizon Designation

The five major master diagnostic horizons O, A, E, B and C have earlier been discussed in chapter 2. However, when a profile is described, each of the five master horizons could be further subdivided into two, three or more horizons depending on the observed features as shown in Fig. 7.1.

Furthermore, to facilitate a more complete description of horizon characteristics than is possible with the major horizon designations indicated in Fig. 7.1, the following symbols are used as suffices together with the major horizon designation when the need arises as follows:

a.   Highly decomposed organic material. This symbol is used with master horizon "O" to indicate the most highly decomposed of the layer(s) with organic materials e.g. Oa

b.   Buried genetic horizon e.g. Ab

c.   Accumulation of concretions or hard non-concretionary nodules. This symbol is not used if concretions or nodules are $CaCO_3$ or more soluble salts, but is used only if concretions or nodules are Fe, Mn, Al or Ti.

d.   Physical root restriction such as plough pans or mechanically compacted zones.

e.   Organic material of intermediate decomposition. It is often used with 'O' e.g. Oe.

f.   Frozen soil – horizon containing permanent ice or permafrost; usually common in very cold regions like Alaska, U.S.A.

g.   Strong gleying, usually due to reduction caused by stagnant water for a prolonged period of time. Chroma of strongly gleyed soils is often ≤ 2.

h.   Illuvial accumulation or organic matter, usually in B horizons of soils in which organic matter-sesquoxide complex have been eluviated from overlying horizons e.g. Bh

i.   Slightly decomposed organic material e.g. Oi.

k.   Accumulation of carbonates. This symbol is used to indicate accumulation of alkaline earth carbonates which effervesces with dilute HCl e.g. Bk or Ck

m.   Indicates continuous or nearly continuous cementation. Roots penetrate "m" horizons only through cracks.

n.   Accumulation of sodium (exchangeable). In such horizons exchangeable sodium percentage (ESP) is often greater than 15%.

o.   Residual accumulation of sesquoxides.

p.   Disturbance due to ploughing, cultivation or animal traffic. A disturbed organic horizon is designated Op. A disturbed mineral horizon even though clearly once an E, B or C horizon is designated Ap.

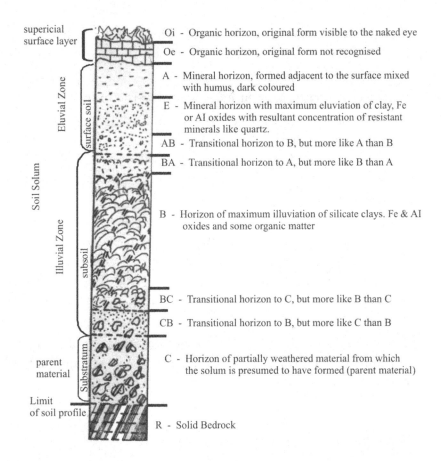

superical surface layer
Oi - Organic horizon, original form visible to the naked eye
Oe - Organic horizon, original form not recognised

A - Mineral horizon, formed adjacent to the surface mixed with humus, dark coloured

E - Mineral horizon with maximum eluviation of clay, Fe or Al oxides with resultant concentration of resistant minerals like quartz.

AB - Transitional horizon to B, but more like A than B

BA - Transitional horizon to A, but more like B than A

B - Horizon of maximum illuviation of silicate clays. Fe & Al oxides and some organic matter

BC - Transitional horizon to C, but more like B than C

CB - Transitional horizon to B, but more like C than B

C - Horizon of partially weathered material from which the solum is presumed to have formed (parent material)

R - Solid Bedrock

Eluvial Zone

Illuvial Zone

Soil Solum

surface soil

subsoil

Substratum

parent material

Limit of soil profile

Fig. 7.1: *Hypothetical mineral soil profile showing the major horizons that may be present*

r.   Weathered or soft bedrock. The symbol is often used with "C" to indicate layers of soft bedrock material which can be dug with a spade such as Cr.

s.   Illuvial accumulation of sesquioxides, such as "Bhs" which indicates the accumulation of illuvial, amorphous, dispersable organic matter – sesquioxides complexes.

ss.   Presence of slickensides, an indication of vertic properties.

t.   Accumulation of silicate clay. This symbol is used to indicate an accumulation of silicate clay that either has formed in the horizon or has been moved into it by illuviation. The clay can be in the form of coatings on ped surfaces or in pores or bridges between mineral grains.

v.   Accumulation of plinthite. Plinthite is a non-indurated mixture of Fe & Al oxides, clay, quartz and other diluents that commonly occurs as red soil mottles usually arranged in platy, polygonal or reticulate patterns. Plinthite changes irreversibly to ironstone hardpans or irregular ironstone aggregates on exposure to repeated cycles of wetting and drying.

w.   This symbol is used with "B" to indicate development of colour or structure or both with little or no apparent illuvial accumulation of material.

x.   Fragipan character. Fragipan is a genetically developed firmness, brittleness or high bulk density often associated with loamy textures in the B horizon.

y.   Accumulation of gypsum.

z.   Accumulation of salts more soluble than gypsum.

Many master horizons and layers that are symbolized by a single capital letter will have one or more lower case letter suffices. **Seldom are more than three suffices needed**. When letter suffices are used, they immediately follow the capital letter. If a surface horizon is disturbed, only "p" is used except where there are surface accumulations of $CaCO_3$, $CaSO_4$ or more soluble salts.
When more than one suffix is needed, the following letters, if used, are written first: a, e, i, h, r, s, t and w. Except for Bhs or Crt, none of these letters are used in combination in a single horizon.

A horizon is never designated Bth, Bts or Btw, though a Bw, Bs, or Bh horizon may be above or below a Bt horizon. If a horizon is buried, the suffix "b" is written last. Suffix "b" is used only for buried mineral soils. If more than one suffix is needed and the horizon is not buried, these symbols, if used, are written last: a, d, f, g, m, v, and x, e.g. Btc, Btg, Ckm and Bsv.

Lower case letter suffices are not used with transitional horizons unless needed for explanatory purposes; for example, use of "k" is appropriate in the sequence A-ACkl-ACk2-AC-C to indicate an accumulation of carbonates in the upper parts of the AC horizon. Commonly, a horizon or layer designated by a single combination of letters needs to be subdivided. The Arabic numerals used for this purpose always follow all letters. Within a C, for example, successive layers could be C1, C2, C3, etc; or if the lower parts is gleyed and the upper part is not, the designations could be C1-C2-Cg1 - Cg2 - R.

One other feature commonly observed during the description of a soil profile is the occurrence of **Lithologic Discontinuities**. A discontinuity is a significant change in particle-size distribution or morphology that indicates a difference in the material from which the horizons formed or, except for some buried soils, a significant difference in age. Arabic numerals are used as prefixes to indicate discontinuities and they precede A, E, B, C and R. Stratification common to soils formed in alluvium is not designated as discontinuities even if particle-size distribution differs markedly from layer to layer unless genetic horizons have formed in the contrasting layers.

Where a soil has formed entirely in one kind of parent material, a prefix is omitted from the symbol; the whole profile is in one material. Similarly, the upper-most material in a profile having two or more contrasting materials is understood to be material "1", but the number is omitted. Numbering starts with the second layer of contrasting material, which is designated "2". Underlying contrasting layers are numbered consecutively. Even though a layer below material "2" is similar to material "1", it is designated "3" in the sequence. The numbers indicate a change in the material, not the type of material. Where two or three consecutive horizons formed in one kind of material, the same prefix number is applied to all the horizon designations in that material e.g. Ap-E-Btl-2Bt2-2BC.

Soil Colour

One of the most apparent characteristic of a soil profile to a layman is its pattern of colours. Colour serves as the first criterion in the delineation of horizons within a given soil profile.

Although colour has little direct influence on the functioning of the soil, one may infer a great deal about a soil from its colour, as in the following examples:

(i)   The content of organic matter in soil is approximately indicated by the darkness of the soil. This characteristic is especially common in the surface soil.

(ii) A bright-coloured subsoil indicates good drainage whereas dull colours indicate poor drainage.

(iii) Soils formed in well oxidized, iron-enriched parent material are usually reddish or brightly coloured.

(iv) Imperfectly and poorly drained soils are nearly always mottled with various shades of grey, brown and yellow, especially within the zone of fluctuation of the water table. The word "mottled" means marked with spots of colour.

Soil colour is a combination of three factors:

(1)   *Hue,* corresponding to the spectral colours, such as red, yellow, blue and green, which is a measure of the dominant wavelength of light.

(2)   *Value* (sometimes called brilliance), the relative lightness or darkness of colour which is a measure of the total quantity of light reflected and

(3)   *Chroma,* the relative purity of the particular spectral colour or the degree of vividness in contrast to greyness, which is a measure of the dominance of one wavelength. It increases as greyness decreases.

Determination of Soil Colour

The "Munsell Soil Color Charts" is used in determining soil colour. Soil colour under the Munsell scheme is determined by using a set of standard colours in the form of small mounted chips with notations for hue, value and chroma – the three simple variables that combine to give all colours.

In the soil colour chart, all colours on a given card are of a constant Hue, designated by symbols such as: 10YR; 5YR or 2.5Y, etc. in the upper right-hand corner of the card. Vertically, the colours become successively lighter

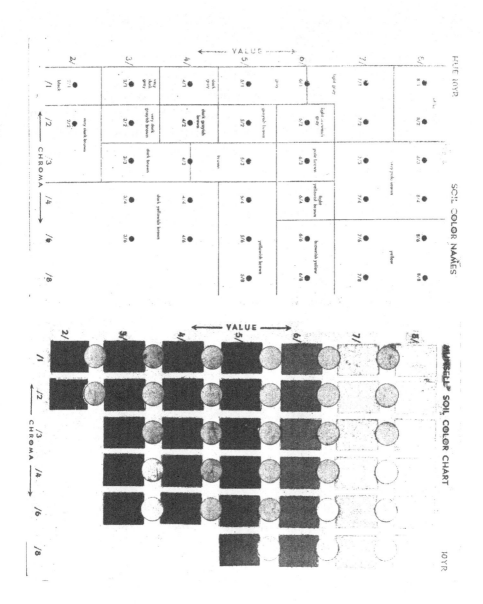

Fig. 7.2: *Soil colour names for several combinations of value and chroma and hue 10YR.*

by visually equal steps; their value increases. Horizontally, they increase in Chroma to the right and become greyer to the left. The Value and Chroma of each colour chip in the chart are printed on the extreme left and at the bottom of the page respectively. Opposite each page or card of colour chips is a page of colour symbols and corresponding English names (Fig. 7.2). An example of a colour notation made for a soil horizon colour is 10YR 6/3. The interpretation of the notation is 10YR (10 Yellow-Red) Hue, a Value of 6 and a Chroma of 3. It is always written in that order, with a slight space between the Hue and the Value/chroma. The proper name for this colour, which is shown on the adjacent left hand page is pale brown, and the correct and final presentation of this colour notation when describing colour is pale brown (10YR 6/3).

It is pertinent to add that soil colour is best determined in sunlight with the light coming over the shoulder. Also, the moisture-status at the time the soil colour is described should always be indicated because dry soil is usually about 2 units higher in Value than the same soil moist. There may also be a difference of 1 unit in the dry and moist chromas of a soil, while the hue generally remains the same.

Mottling

The word "mottled" means marked with spots of colour. Mottling in soils is described by noting:

(1)   the colour of the matrix and the colour or colours of the principal mottles and
(2)   the pattern of the mottling.

The colour of the mottles may be defined by using the Munsell notation as already described. The pattern of mottles may be described by three sets of notations: abundance, size and contrast. Abundance of mottles is indicated in three general classes as: few (mottles less than 2% of surface area of horizon), common (2-20% of surface area) and many (>20% of surface area). This is based upon the relative amount of mottled surface in the unit area of the exposed soil horizon. Size refers to the approximate diameters of individual mottles as follows: fine (<5mm), medium (5-15mm) and coarse(>15mm). Contrast between mottles may be described as faint(hue and chroma of matrix and mottles are similar; distinct (matrix and mottles vary 1-2 hues and several units in chroma and value) and prominent (matrix and mottles vary several units in hue, value and chroma). A full description of a typical mottle in a soil horizon would be common, medium, distinct red (2.5YR 4/8) mottles.

In general, the significance of the occurrence of mottles is the fact that they indicate, where they occur, that a soil has had periods of inadequate aeration or reduction during some period of the year, leading to cycles of oxidation-reduction reactions.

Soil Texture

Mineral soils are often composed of inorganic particles of varying sizes called soil separates. Soil texture refers to the relative proportions of the various separates or size groups of individual soil grains in a mass of soil. Specifically, it refers to the proportions of sand, silt and clay below 2000μm (2mm) in diameter in a mass of soil.

In general, sand is coarse and gritty, silt is smooth like flour and clay is sticky and plastic when wet. A soil that exhibits a combination of the three properties is said to have a loam texture. Sandy textures can be modified by designating the "coarseness" or "finess" of the dominant size of sand particles e.g. coarse sandy loam or fine sandy loam. The presence of coarse particles larger than 2000μm in diameter but smaller than 2,500 um (25mm) is also recognized by modifiers of textural class names like "gravelly sandy loam" or "cobbly loam".

Rarely does a soil sample consist wholly of one soil separate; so combinations of at least two separates give rise to classes of soil texture. Two schemes of classification used for defining various separates in soil include those of the United States Department of Agriculture (USDA) and the International Soil Science Society(ISSS) proposed by Atterberg. These two schemes are presented in Table 7.1. Note the overlap in the size ranges for the sand and silt fractions in the two schemes. While the sand fraction in the USDA scheme ranges from 2000-50μm, it ranges between 2000-20μm in the ISSS scheme. It is therefore pertinent to always indicate the scheme of classification used when the texture of a soil is determined.

Soil texture is studied because it is related to certain physical properties of soil such as plasticity, permeability, ease of tillage, fertility, water holding capacity and overall soil productivity. For instance, for irrigation purposes, loamy and clay textures are classed as soils of high moisture holding capacity while loamy sands and sands have low moisture holding ability. Texture also determines the microbiological population of a soil and hence the biological and biochemical reactions taking place in such a soil.

Table 7.1: *Size limits of soil separates in the USDA and ISSS schemes*

USDA Scheme		ISSS Scheme	
Name of Separate	Diameter Range (μm)	Name of Separate	Diameter Range (μm)
Very coarse sand	2000-1000		
Coarse sand	1000- 500	Coarse sand	2000 - 200
Medium sand	500-250		
Fine sand	250-100	Fine sand	200 - 20
Very fine sand	100- 50		
Silt	50- 2	Silt	20 – 2
Clay	< 2	Clay	< 2
Coarse Fragments			
Gravels	2000-7500μm	(2-75mm)	
Cobbles	75000-254000μm	(75-254mm)	
Stones	>254000μm	(>254mm)	

Definition of Soil Textural Classes

Apart from the modification of sandy textures with terms such as gravelly, coarse, very fine, fine, etc., there are twelve basic soil textural classes recognized. In order of increasing proportions of the fine separates, the classes include: sand, loamy sand, sandy loam, loam, silt, silt loam, sandy clay loam, clay loam, silty clay loam, sandy clay, silty clay and clay.

The basic soil textural class names in present use are defined in terms of particle-size distribution as determined in the laboratory by a procedure termed Particle Size Distribution Analysis or Mechanical Analysis or Granometric Analysis. The percentage of size fractions combined to determine the texture using a standard soil textural triangle is shown in Fig. 7.3. In general, the twelve textural classes may be defined as follows:

Sand   -   Soil material that contains 85% or more of sand and a percentage of silt plus 1½ times the percentages of clay not exceeding 15.

Loamy Sand   -   Soil material that contains at the upper limit 85 to 90% sand, and the percentage of silt plus 1½ times the percentage of clay is not less than 15; at the lower limit it contains not less than 70 to 85% sand, and the percentage of silt plus twice the percentage of clay does not exceed 30.

Sandy Loam - Soil material that contains either 20% clay or less and the percentage of silt plus twice the percentage of clay exceeds 30, and 52% or more sand; or less than 7% clay, less than 50% silt, and between 43% and 52% sand.

Loam - Soil material that contains 7 to 27% clay, 28 to 50% silt, and less than 52% sand.

Silt loam - Soil material that contains 50% or more silt and 12 to 27% clay (or) 50 to 80 percent silt and less than 12% clay.

Silt - Soil material that contains 80% or more silt and less than 12% clay.

Sand clay loam- Soil material that contains 20 to 35% clay, less than 28% silt and 45% or more sand.

Clay loam - Soil material that contains 27 to 40% clay and 20 to 45% sand.

Silty clay loam - Soil material that contains 27 to 40% clay and less than 20% sand.

Sand clay - Soil material that contains 35% or more clay and 45% or more sand.

Silty clay - Soil material that contains 40% or more clay and 40% or more silt.

Clay - Soil material that contains 40% or more clay, less than 45% sand, and less than 40% silt.

In the field however, a method known as the "Feel Method" is used in assessing soil texture. In this method, a sample of the soil usually moist or wet is rubbed between the fingers and the texture assessed by the behaviour of the soil particles using a knowledge of the behaviour of the various quantities of the separate present in the soil sample. In other words, the differing sizes of the constituent particles give each soil a characteristic feel. Thus, a soil composed mainly of coarse sand particles feels light and gritty;one that is composed mainly of silt feels smooth or powdery when dry and floury when wet while one composed mainly of clay feels heavy and sticky when wet.

In general, the twelve textural class names already established, form a more or less graduated sequence from soils that are coarse in texture and easy to handle to the clays that are very fine and difficult to manage, while the loams are in between the two extremes as outlined in Table 7.2.

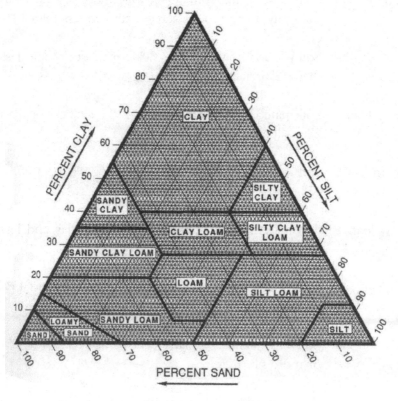

Fig. 7.3: *Standard USDA Soil Textural Triangle*

Table 7.2: *Classification of Soil Textural Classes (USDA System)*

General Terms		Basic Soil Textural
Common Names	**Texture**	**Class Names**
Sandy soils	Coarse.........	{ Sand / Loamy sand
	Moderately coarse	{ Sandy loam / Fine sandy loam
Loamy soils	Medium	{ Very fine sandy loam / Loam / Silt / Silt loam
	Moderately fine	{ Clay loam / Sandy Clay loam / Silty clay loam
Clayey soils	Fine	{ Sandy clay / Silty clay / Clay

Adapted from Brady (1974).

Soil Structure

Soil structure may be defined as the aggregation of primary soil particles (sand, silt and clay) into compound particles termed peds or aggregates, which are separated from adjoining peds by lines of weakness.

Field description of soil structure is split into three namely; Grade, Class and Type. In other words, field terminology for structure consists of separate set of terms designating each of the three qualities, which by combination form the names for structure. Generally, the *grade* of structure is written first, followed by the *class* and then the *type* of structure, in that order.

Grade of soil structure refers to the degree of aggregation and expresses the differential between cohesion within aggregates and adhesion between aggregates. In field practice, grade of structure is determined mainly by noting the resistance of the aggregates when they are gently crushed. Grade of structure is, therefore, described by such terms as weak, moderate, strong or structureless. A structureless condition exists when there is no observable degree of aggregation or no definite orderly arrangement of natural lines of weakness. If a structureless soil is coherent, it is termed massive. This grade of soil structure exists in medium to fine textured soils that have been puddled. If the structureless soil is incoherent, it is described as single grained. This is the grade of soil structure that typically exists in coarse-textured soils, such as sands.

Class of soil structure refers to the size of the aggregates or ped and is described as very fine, fine, medium, coarse and very coarse. Type   of soil structure refers to the shape of the aggregates or peds and includes such forms as granular, crumb, platy, blocky, subangular blocky, prismatic and columnar. The description and common location within profiles of various types of structure is presented in Table 7.3, while Table 7.4 shows the precise limits for the description of classes for each type of soil structure.

Thus, the full description of the structure of a given ped could be either strong, coarse, prismatic structure; moderate fine granular structure; weak fine crumb structure or structureless (massive) or structureless (single-grained).

Soil structure is important in agriculture from the point of view that a well aggregated soil is often well drained, has good permeability of water, air and roots. Such a soil is also easily worked or tilled and thus serve to control erosion. All these are made possible because of the numerous macro pore spaces created by the existence of numerous lines of weakness between aggregates or peds.

Common agents of aggregation which are responsible for binding primary soil particles into peds include the following:
(i)   Colloidal clay minerals which consist of the finer, more reactive part of clay in soils.
(ii)   Colloidal oxides of iron, aluminum and manganese, which are collectively termed sesquioxides. These are especially typical of tropical soils.
(iii) Microbial gums; which are gums secreted by micro-organisms in soils.
(iv)   Organic compounds, especially humus which are also colloidal in nature
(v)   Carbonates.

Table 7:3: *Diagrammatic Definition and Location of Various Types of Soil Structure*

Structure Type	Aggregate Description	Diagrammatic Aggregate	Common Horizon Location
Granular	Relatively nonporous, small and spheroidal peds, not fitted to adjoining aggregates		A horizon
Crumb	Relatively porous, small and spheroidal peds; not fitted to adjoining aggregates		A horizon
Platy	Aggregates are plate like, Plates often overlap and Impair permeability		E horizon forest and clay pan soil
Blocky	Block like peds bounded by other aggregates whose sharp angular faces form the cast for the ped. The aggregates often break into small blocky		Bt horizon
Subangular Blocky	Block like peds bounded by other aggregates whose rounded subangular faces form the cast for the ped		Bt horizon
Prismatic	Column like peds without rounded caps. Other prismatic aggregates form the cast for the ped. Some prismatic aggregates break into smaller blocky peds		B horizon with expand-ing clays
Columnar	Column like peds with rounded caps bounded laterally by other columnar aggregates that form the cast for the peds		B horizon with non-expanding and/or expanding silicate clays

Adapted from Soils Laboratory Exercise Source Book, Am. Soc. Of Agron. 1964.

Table 7.4: *Types and Description of Soil Structure*

Spheroidal			
Characteristic of Surface (A) horizons. Subject to wide and Rapid changes.	Granular (porous) Crumb (very porous)		(a)
Plate-like			
Common in E-horizons, may occur in any part of the profile. Often inherited from parent material of soil, or caused by compaction.			(b)
Block-like	Angular blocky		(c)
Common in B-horizons, particularly in humid regions. May occur in A-horizons.			
	Subangular blocky		(d)
Prism-like	Columnar (rounded tops)		(e)
Usually found in B-horizons. Most common in soils of arid and semi-arid Regions.			
	Prismatic (flat, angular tops)		

Soil Consistence

Soil consistence is a term used to describe the degree of cohesion and adhesion or resistance to deformation or rupture between soil peds or aggregates. It is also a measure of the physical condition of a soil at various moisture contents, as evidenced by the behaviour of that soil towards mechanical stresses or manipulations. The stress is commonly evidenced by feeling the soil, manipulating it by hand or by pulling a tillage implement through it.

Whereas soil structure deals with the shape, size and distinctness of natural soil aggregates, consistence deals with the strength and nature of the forces between particles. Consistence is thus important for tillage and traffic considerations. Hence, sandy soils exhibit minimal cohesive and adhesive properties and is so easily deformed that automobiles easily get stuck. On the other hand, clay soils can be so sticky when wet as to make hoeing or ploughing difficult.

Consistence is described at three soil moisture levels: wet, moist and dry. The terms used to describe consistence include:

(i)   Wet soil – often described in terms of stickiness as nonsticky, slightly sticky, sticky, very sticky; and in terms of plasticity as non plastic, slightly plastic, plastic, and very plastic.

(ii)   Moist soil – this is very important because it best describes the condition of soils when they are tilled in the field. Consistence of a moist soil is described in the following terms; going from the material with least coherence to that which adheres so strongly as to resist crushing between the thumb and forefinger: loose, very friable, friable, firm, very firm and extremely firm.

(iii) Dry soil – terms used to describe the degree of rigidity or brittleness to crushing or manipulations include the following: loose, soft, slightly hard, hard, very hard and extremely hard.

Cementation is also a type of consistence and is caused by cementing agents such as calcium carbonate, silica or oxides of iron and aluminum. Cementation is little affected by moisture content. The terms used to describe cementation include: weakly cemented i.e. cemented units can be broken in the hand. Strongly cemented – units cannot be broken in the hand but can be broken easily with a hammer. Indurated – units breakable only with sharp blows of a hammer.

Cutans

Cutans may be defined as modifications of the fabric of natural surfaces in soil materials due to concentration of particular solid constituents such as clay, sesquioxides and organic materials notably humus.

The presence of cutans in subsoil horizons (B or C) of a soil profile is of specific pedological significance. For instance, the presence of recognizable amounts of clay cutans or clay skins on ped faces or in pores is a direct indication that an argillic horizon (Bt) is present and that such a soil is mature. The presence of both Fe-oxyhydroxide (sesquioxides) and humus (organs) in the B-horizon may indicate the occurrence of a spodic horizon and the onset of the process of podzolisation in a soil.

Cutans are often detected on ped faces, soil pores or within cracks or channels as coating (skins) or bridges between sand grains. They are often described in the field (best with 10 x hand lens) in terms of their quantity (very few, few, common, many and continuous); thickness (thin, moderately thick and thick) and location (ped faces, pores, bridges). Cutans occur in soils chiefly as a result of eluviation-illuviation processes.

Soil Pores

Pores are spaces or voids between solid soil materials. The occurrence or abundance of pores in soils is of pedological significance because a soil with many coarse pores will be much more aerated and better drained than one with few very fine pores.

When describing pores in detail, a definite sequence of terms should be consistently followed. The usual sequence is *number* (few, common, many), *size* (very fine, fine, medium, coarse), *continuity* (discontinuous, constricted, continuous), *orientation* (vertical, horizontal, random, oblique), *shape* (vesicular, irregular, tubular) and location (imped, exped).

A typical description of peds in a particular horizon would be "many, very fine, continuous, vertical, simple, tabular pores imped". Generally, however, pores are described mainly by noting the number, size and perhaps their continuity as in the example "many fine continuous pores" or "many coarse pores".

Roots

By recording the presence or absence of roots in horizon descriptions, the pedologist accumulates direct evidence to substantiate appraisal of the effects of soil properties on the behaviour of plant life in the soil. This is in the sense that plant roots fail to penetrate some soil horizons because of deficiencies of moisture, nutrients, oxygen or because of extremely unfavorable physical or chemical properties. Ordinarily, roots cannot penetrate cemented hardpan layers unless the pans are fractured.

A large part of the roots of most plants is in the upper soil horizons. For most grasses, 65 to 80% of the roots are in the upper 15cm. The roots of many forest trees are concentrated in the upper soil horizons and total root volume decreases as depth increases. If soil moisture and temperature conditions are favourable, some grass roots can penetrate to depths of 3m or more. Also, maize plants have been known to remove available moisture to depths of 1.5m or more, while roots of oil palm can go even deeper than 7.5m (Soil Survey Staff, 1975).

Roots in horizons are often described in terms of the *number* (few, common, many) and *size* (very fine, fine, medium, coarse). Thus we can have "many very fine roots"," common coarse roots", etc.

Included Materials

Inclusions often observed in the horizons of soil profiles which could be of pedological significance include concretions, nodules, stones, weathered rock pieces, minerals, charcoal, pottery, fecal pellets or faunal excreta, etc.

Local concentrations of oxides of Fe, Mn and Carbonates ($CaCO_3$) constitute concretions and nodules in soils. Nodules are often uniform concentrations of the oxides or carbonates while concretions are often concentric rings or admixtures of usually two or more materials say Fe and Mn. Thus, we have Fe or Mn or $CaCO_3$ nodules and Fe-Mn or Mn-$CaCO_3$ concretions.

The presence of charcoal and pottery pieces often suggests the existence of old settlements in the past. While rock and mineral pieces suggest the nature or origin of parent material from which a soil may have developed. Whether the minerals present are resistant or easily weatherable could also give some insight into the weathering potential of a soil.

Descriptions of included materials should include information on abundance, size, hardness, shape, colour and nature of the materials(s). thus, typical examples could be; many, medium, roundish basaltic Fe-Mn concretions; common fine mica flakes; few medium ant holes, etc.

Horizon Boundary Characteristics

Horizon boundaries within a given soil profile vary in *distinctness* and in surface *topography*. Some boundaries are clear and sharp, as those between A and B horizons. Again, they may be diffuse, with one horizon gradually merging into another as in transitional AB and BA horizons. The distinctness of the horizons to the observer depends partly upon the contrast between them and partly upon the width of the boundary itself.

The characteristic widths of boundaries between soil horizons may be described as (1) *abrupt*, if less than 25mm wide (2) *clear*, if about 25 to 60mm wide (3) *gradual*, if 60 to 120mm wide; and (4) *diffuse*, if more than 120mm wide.

The topography of different soil horizons varies as well as their distinctness. Horizon boundaries may thus be described as (1) *smooth*, if nearly a plane; (2) *wavy*, if pockets are wider than their depth; (3) *irregular*, if irregular pockets are deeper than their width (4) *broken*, if parts of the horizon are unconnected with other parts.

Example of Soil Profile Description

In order to facilitate the collection of all the necessary information at a soil profile site, a "Soil Description Sheet" which incorporates both the general site information and the horizonal description as in the example shown in Fig. 7.4 is often taken to the field. The relevant information required are then made and recorded in the sheet.

In the office, the "short-hand" field descriptions are transcribed into the "long-hand" description which is incorporated in the soil report as in the example shown in Fig. 7.5. Note that the sequence of recording of the various parameters in the "long-hand" is the same as in the Soil Description Sheet used in the field.

(a)   Site Information

Described by ....

Date...

Location ....

Soil Mapping Unit ....

Taxonomic Classification ....

Elevation ....

Climate: MAR
          Rainy Period::
     MAT:...
          Soil Moisture Conditions:......................................
          Soil temperature conditions:...................................

Soil Parent Material ....

Geology ....

Geomorphology & Physiography ....

Topography ....

Local relief:: Crest, Upper slope; Mid-slope; Lower slope; valley floor; Back swamp; Levee

*Slope:*Level or nearly level 0-2%, gently sloping or undulating 2-4%
     Moderately to strongly sloping 4-7%; very strongly sloping 7-12%;
     Steeply sloping and hilly 12-18%; very steep or hilly 18-30%.

Vegetation:
               General:...
               Land Use: ....
               Uncultivated spots:................................
Erosion: Sheet – none, slight, moderate, severe
               Gully – none, slight, moderate, severe
               Deposition: ....
Drainage: V. poorly dr.; poorly dr.; mod. well dr., well dr.;
               V.well dr.; somewhat excess dr.; excess drained.

Depth to Water Table:..

Depth to Impenetrable Layer:...

Figure: 7.4: Soil Description sheet used in the field

(b) Horizonal Descriptions

Horizon Depth (cm)	Colour	Texture	Mottles	Structure	Consistence	Cutans	Pores	Miscellaneous Observations	Boundary
	Moist/ Dry **Name** .... **Notation** 10R........ 2.5YR...... 5YR......... 7.5YR...... 10YR....... 2.5Y........ 5Y	stony cobbly gravelly sand loamy sand sandy loam loam silt loam silt sandy clay loam clay loam silty clay loam sand clay silty clay Clay	**Munsell Colour** Name .... Not ...... **Abundance** Few (1-2%) Common (2-20%) Many (20%) **Size** Fine (5mm) Medium(5-15mm) Coarse(>15mm) **Contrast** Faint, distinct prominent	**Grade** weak, moderate strong **Class** fine, medium coarse **Type** crumb, granular sbk, abk. prismatic columnar platy **structureless** single grained Massive	**Wet** Non-sticky, slightly stickyk V. sticky **Moist** Loose, very friable, firm very firm **Dry** Loose, soft, slightly hard hard, very hard	Type.......... **Frequency** V. few (5%) f. (5-25%) c (25-50%) m (50-90%) **continuous** (90%) thickness thin, mod, thick **location** ped faces, pores, bridges	**Abundance** f, c, m. **size** V f, f, M, c. **Roots** **Abundance** f, c, m. **Size** v. f, (0.1mm) v, (1-2mm) m, (2-5mm) coarse (2.5mm)	*Concretions/* *Nodules* .... *Minerals* .... *Animal* *Activities* .... *Reaction* .... **Others**	**Distinctness** Abrupt Clear Gradual Diffuse **Topography** Smooth Wavy Irregular broken
	Moist/ Dry **Name** .... **Notation** 10R........ 2.5YR...... 5YR......... 7.5YR...... 10YR....... 2.5Y........ 5Y	stony cobbly gravelly sand loamy sand sandy loam loam silt loam silt sand clay loam clay loam silty clay loam sandy clay silty clay Clay	Munsell Colour Name .... Not ...... **Abundance** Few(1-2%) Common (2-20%) Many (20%) **Size** Fine (5mm) Medium(5-15mm) Coarse(>15mm) **Contrast** Faint, distinct prominent	**Grade** weak, moderate strong **Class** fine, medium coarse **Type** crumb, granular sbk, abk. prismatic columnar platy **structureless** single grained Massive	**Wet** Non-sticky, slightly stickyk V. sticky **Moist** Loose, very friable, firm very firm **Dry** Loose, soft, slightly hard hard, very hard	Type.......... **Frequency** V. few (5%) f. (5-25%) c (25-50%) m (50-90%) **continuous** (90%) thickness thin, mod, thick location ped faces, pores, bridges	**Abundance** f, c, m. size V, f, f, m c. **Roots** Abundance f, c, m. **Size** v. f, (0.1mm) v, (1-2mm) m, (2-5mm) coarse (2.5mm)	*Concretions/* *Nodules* .... *Minerals* .... *Animal* *Activities* .... *Reaction* .... **Others**	**Distinctness** abrupt clear gradual diffuse **Topography** smooth wavy irregular Broken

Figure: 7.5: *"Long-hand" Profile Description of Kukum Daji Series (KD)*

Profile Nr.	KD-4
General Site Information	
Location	: Traverse 3 point 2
Taxonomic classification	: Oxic Haplustult (USDA) Orthic Acrisol (FAO)
Soil parent material	: Weathered basaltic lava.
Geology	: Basaltic lava over Basement Complex.
Geomorphology	: Nearly level plain within an Inselberg landscape of the Kagoro hills
Topography	: Valley floor; nearly level, 1% slope.
Vegetation/Landuse	: Cultivated with irrigated maize, vegetables and orchard.
Soil erosion hazard	: None encountered
Drainage	: Well drained
Depth to water table	: None encountered
Depth to impenetrable layer	: None encountered at 200cm when augred.
Date sampled	: April 17, 1990
Horizon Descriptions	

Horizon	Depth (cm)	Description
Ap	0-16	Dark yellowish brown (10YR 4/4) loam; moderate medium subangular blocky structure; sticky and slightly plastic (wet) friable (moist) slightly hard (dry); many medium pores; common very fine mica flakes; common medium ant holes; clear smooth boundary.
Bt	16-98	Strong brown (10YR 5/6); moderate medium subangular blocky structure; sticky & plastic (wet) friable (moist); few thin clay cutans on ped faces; common medium pores; common medium and coarse roots; few medium sub-rounded basaltic Fe-Mn concretions; many fine mica flakes; few medium ant holes; gradual smooth boundary.

Btc	98-160	Strong brown (10YR 5/6) gravelly clay with many medium distinct light yellowish brown (10YR 6/4) and red (2.5YR 4/6) mottles; moderate medium subangular blocky structure; continuous moderately thick clay and Fe-oxyhydroxide cutans on ped faces: common fine pores; common medium roots; many medium roundish basaltic Fe-Mn concretions; common fine mica flakes.

Revision Questions

1. (a) Outline the broad criteria often used by pedologists in delineating soil bodies in the field.
    (b) What precautions must you take to ensure the proper characterization of a soil body in the field?

2. (a) List ten morphological characteristics that are taken into consideration when a soil profile is described in the field.
    (b) With the aid of a well labeled diagram *only*, show the detail sequence of horizonation within a hypothetical mineral soil profile.

3. (a) Outline four possible inferences that can be drawn from a knowledge of soil colour.
    (b) Distinguish between mottling and strong gleying in soils. How are mottles characterized in soils?

4. (a) Discuss soil texture in relation to the water holding capacity of soils.
    (b) List the factors which promote the development of a good soil structure.

References

Brady, N. C. 1974. *The nature and properties of soils*. 8th Edition. Macmillan Publishing Co., Inc., New York, USA. 639pp.

FAO. 1977. *Guidelines for soil profile description.* 2nd Edition. FAO, Rome. 66pp.

Schoeneberger, P. J., D. A. Wysocki, E. C. Benham and W. D. Broderson(eds). 2002. Field book for describing and sampling soils. Version 2.0. NRCS-USDA, National Soil Survey Center, Lincoln, NE. 1-1: 9-13.

Soil Survey Staff. 2006. *Soil Survey Manual.* U.S. Dept. Agric. Handbook 18. U.S. Govt. Printing Office, Washington.

Soil Survey Staff, 1975. Soil Taxonomy: A basic system of soil classification for making and interpreting soil surveyors. USDA Agric: *Handbook 436.* US Govt. Printing Office, Washing ton D.C. 754pp.

8

Laboratory Data Used in Pedological Studies

Introduction

Laboratory data are very critical to the understanding of the properties, processes and genesis of a soil pedon and are thus a *sine qua non* to understanding the classification, use and management of soils. Eventhough experienced pedologists can make a great deal of inferences from the morphology they see and describe in the field and can make predictions about the properties and qualities of a soil, far more accurate interpretations and predictions about soil quality for modern agriculture and nonfarm uses as well as the accurate taxonomic and technical classification of soils can be made if quantitative laboratory data are available.

All modern soil classification systems place a great deal of reliance on information about the quantitative properties of soils. Certain kinds of soil properties are selected for use in the classification process based on their assumed importance in understanding the genesis of the soil and the number of other important properties that co-vary with changes in the property under consideration (Buol et al., 1997).

Pedon Characterization Data

A number of selected soil physical, chemical, mineralogical and micro morphological data are often associated with pedological soil characterization studies. Such properties and brief mention of the methods used for their determinations include the following:

Physical Properties
- *Particle Size Distribution Analysis* – involves the measurement of the

percentage of the mineral separates; sand, silt, and clay. The procedure has two aspects; dispersion of the soil and fractionation into particle size classes. Dispersion is achieved by shaking or stirring the soil sample with a sodium hexametaphosphate (calgon) solution. Determination of the content of silt and clay is by a sedimentation-Pipette or sedimentation-hydrometer procedure, while the sand fraction is determined by sieving (a nest of sieves is used when sand sub-fraction percentages are to be measured) as outlined by Day (1965) and SCS-USDA (1972).

- *Bulk Density* – is measured as the *weight per unit volume* of a soil and expressed in units of megagram per cubic metre (Mg m^{-3}) or gram per cubic centimeter (g cm^{-3}) in non-SI unit. Methods for determining bulk density data are the undisturbed core, plastic or paraffin-coated clod, excavation and gamma radiation densitometry techniques (Blake, 1965). By and large the core method using undisturbed core samplers of known volume is more popular in pedological studies involving many soil horizons. The clod method is used in collecting data from soil horizons too dense or brittle to sample with a coring device, and it also allows calculation of shrink-swell potential or linear extensibility from the same clod specimens (Grossman et al., 1968).

- *Coefficient of Linear Extensibility (COLE) or Shrink-Swell Capacity* – is a derived value. It is computed from the difference in bulk density between a moist clod and an oven dry clod. It is based on the shrinkage of a natural soil clod between a water content of 33kPa(10kPa for sandier soils) and 1500 kPa i.e. field moisture capacity and oven dry (1500kPa) soil. COLE is defined as follows:

$$COLE = \frac{L_m - 1}{L_d}$$

Where L_m = Length of moist sample
L_d = length of dry sample

The coefficient is actually calculated from differences in bulk density of coated clods or undisturbed core samples when moist (33kPa or 10kPa if coarse sandy soil) and oven dry as:

$$COLE = \sqrt[3]{\frac{Dbd}{Dbm}} - 1$$

Where Dbd = bulk density dry

      Dbm = bulk density moist

- *Soil Moisture Content* – of interest and use in soil genesis and characterization is the percent (volume basis) of water retained at 1500 kPa, as this is the "permanent wilting point" for plants and refers to "dry" soil for classification purposes. In general, tension plate and pressure chamber techniques, using soil cores with field structure preserved are used for soil moisture retention measurements at low suction values while pressure membrane apparatus, using crushed soil samples is used for higher tension measurements, especially the 1500 kPa limit (Richards, 1965, SCS-USDA, 1972). Usually, Available Water Capacity or Water Retention Difference is computed from water retentions at 33 kPa or 10 kPa for sandier soils and the 1500 kPa water suction. It is converted to cm of water per cm of depth of soil through use of the bulk density.

Chemical Properties

- *Soil pH*- is measured in water and in salts. The pH measured in water is determined in distilled water typically mixed 1:1 with dry soil. The pH measured in potassium chloride is determined in 1N KC1 solution mixed 1:1 with soil. The pH measured in calcium chloride is determined in 0.01M $CaCl_2$ solution mixed 2:1 with soil.

  Measurements of pH in a dilute salt solution is common because it tends to mask seasonal variations in pH, as the readings remain uniform regardless of the time of the year. Measurements in KCl are more popular in regions with acid soils while $CaCl_2$ should be used in less acid regions. Soil pH may also be measured in 1N NaF. This measurement is usually used to identify soils that are dominated by short-range-order minerals such as in Andisols and Spodosols.

- *Organic Carbon*-The wet digestion method of Walkley-Black (1935) in which the soil is digested in an excess of chromic acid with titration of the unused oxidant is often used especially in the developing world. In the developed world, most laboratories now estimate organic matter content by the loss on ignition (LOI) method in which a soil sample is combusted at 360 to 400°C for several hours or overnight. The weight loss upon ignition is assumed to be proportional to the organic matter content.

- *Total Nitrogen-* is commonly determined by the macro-Kjeldahl method (Bremner, 1965) where such data are desired for soil genesis and classification purposes.
- *Electrical Conductivity of Saturation Extract (EC)* – This measures the accumulation of soluble salts in the soil solution. This is measured by preparing a saturated paste of soil then vacuum filtering to obtain the saturation extract (US Salinity Laboratory Staff, 1954). The amount of soluble salt in the saturation extract is then measured with a Wheatstone bridge-conductivity cell combination. The results are expressed in units of decisiemens per meter (dSm^1) at 25°C (formerly, mmhos cm^{-1}).
- *Extractable (Free) Iron* – A commonly used procedure for its measurement involves reducing the iron with sodium dithionite, chelating it with sodium citrate in a sodium bicarbonate buffer following which it is measured colorimetrically (Mehra & Jackson, 1960).
- *Cation Exchange Capacity* – Three types of CEC determinations are often made and results vary as the pH at which they are determined also vary. The three methods of determination include: (1) CEC by the summation method (buffered at pH 8.2) in which all exchangeable cation species including Al and H or exchange acidity are added. (2) CEC by ammonium saturation displacement method commonly conducted at pH 7.0 (3) Unbuffered or neutral salt CEC whereby only the basic cations are displaced by ammonium saturation plus the KCl extractable Al and is known as ECEC (Effective CEC).
- *Exchangeable Bases and Exchange Acidity* – Methods for determining exchangeable bases involve displacement of the ions from the soil colloids and their measurement in the displaced solutions. They are measured by flame (K and Na) and atomic absorption photometry or by titration techniques (SCS-USDA, 1972).
- *Percentage Base Saturation (PBS)* –

$$PBS = \frac{\text{Sum of Exchangeable Bases (Ca, Mg, K, Na)} \times 100}{\text{Cation Exchange Capacity (CEC)}}$$

A clear distinction must be made as regards which of the pH at which the CEC being used as the denominator was determined. This is because base saturation determined at each of the pH are used in Soil Taxonomy for different interpretation purposes as would be seen later.

- *Extractable P and Phosphate Retention* – The Bray-1 method is widely used as an index of plant-available P in the soil and is designed to remove the easily acid-soluble P, largely Ca phosphates, and a portion of the phosphates of Al and Fe and is most successful in acid soils (Bray and Kurtz, 1945; Olsen and Sommers, 1982). Phosphate retention refers to the percent P retained by soil after equilibration with 1000 mg kg^{-1} P solution for 24 hours.

Mineralogical/Properties

- In general, two broad groups of minerals are recognized in soils; the skeletal minerals or grains which include primary minerals & microcrystalline aggregates and the clay minerals & amorphous fine grained compounds. The skeletal minerals which are largely of sand and silt size are best determined with the aid of Polarizing Optical Microscope fitted with polarizer while the clay minerals are determined with X-ray Diffraction Analysis, Differential Thermal Analysis, Thermogravimetric Analysis, Infrared and Electron Microscopy.

Thin Sections for Soil Micromorphology

- For micromorphological work, blocks of undisturbed soil with orientation marked (as to which was upper side, in place) are collected and brought to the laboratory for binocular microscopic study and thin section preparation. Preparation of thin sections for microscopic observation can take place using many techniques such as those outlined by Brewer (1964); Grossman (1964); Murphy (1986) and Soil Survey Laboratory Staff (1995). Once a thin section has been prepared, observation techniques need to be used to quantify the observation. Modern imaging techniques useful in micromorphological studies have been outlined by Mermut & Norton (1992).

Handling and Preparation of Samples in the Laboratory

It is very important to note that proper handling of soil samples meant for characterization studies is necessary to prevent contamination and to minimize extreme changes in elemental concentrations or pH between the soil in the field and the sample to be analyzed. Handling includes the proper transportation of the soil cores collected from the genetic horizons of pedons. Once in the laboratory, the soils should be air-dried for two to three days, ground and sieved to pass a 2-mm sieve. Each sample is then stored in accurately labeled soil bags. All chemical & mineralogical determinations as well as the particle size distribution analysis are carried out from the sieved samples.

Pedological Inferences from Laboratory Data

A number of inferences important in elucidating soil forming-processes, soil genesis and soil morphology can be drawn from laboratory data and thus help immensely in the proper placement of soil bodies in their appropriate taxa or class as in the following cases:

Particle Size Analysis

In Pedology, particle size analysis is important particularly for knowing the amount of clay, especially the ratio of fine to coarse clay as well as the subfractions of the total sand content. When these values are known, the following inferences can be made:

- Many important diagnostic subsurface horizons especially the argillic, kandic and natric horizons are defined on the basis of subsoil accumulation of clay or the existence of a 'clay bulge' in the B horizon. There must be evidence of clay translocation in an eluviation-illuviation process within a profile for an argillic horizon to form. The fine clay (<0.2μm) to coarse clay (2-0.2μm) ratio is especially useful as fine clay is present in larger amounts in illuvial horizons.
- The existence of lithologic discontinuities are much easily detected on the basis of shifts in sand subfraction percentages or abrupt shifts in silt and clay percentages in adjacent horizons.
- Particle size data are extremely important in the determination of 'particle-size classes' which constitutes a major differentiate at the family level of the USDA Soil Taxonomy.

Bulk Density

Bulk density being the ratio of the dry mass to the total volume of soil particles plus pore spaces in a sample can be used in making the following inferences in soil classification activity:

- Detection of diagnostic horizons (endopedons) associated with high bulk density and brittleness such as the fragipan, duripan and petrocalcic horizons which have bulk densities higher than 1.8 g cm-3

- Detection and quantification of the presence of volcanic ash material and pumice which is an important criterion for identification of 'andic soil properties' necessary for the recognition of the Andosol Soil Order of Soil Taxonomy. Bulk density of less than <0.90 g cm^{-3} indicates volcanic ash in mineral soils low in organic matter.

- Bulk densities of Histosols range from 0.05 to 0.15 g cm^{-3} for fibric and most of the hemic materials (Lynn et al., 1974). For sapric materials, the range is wider but densities >0.25g cm^{-3} are limited to organic soils with $<7\%$ rubbed fibre of which most are cultivated surface soils.

- Bulk density between 1.6 to 1.8 g cm^{-3} may indicate that aeration and water movement are too low for optimum plant growth.

- As seen earlier, bulk density data is used in the calculation of the coefficient of linear extensibility (COLE). It is also used in the calculation of important soil parameters such as total porosity, available water capacity and carbon contents per unit volume of soil, a criterion for classifying some soils.

Coefficient of Linear Extensibility (COLE)

COLE is a derived value that denotes the fractional change in the clod dimension from a dry to a moist state. COLE is used to make inferences about shrink-swell capacity and clay mineralogy which is important for soil physical qualities (large deep cracks in the dry seasons) determination as well as for genetic processes and soil classification. For instance,

- If COLE exceeds 0.03, a significant amount of montmorillonitic clay is present.

- If COLE exceeds 0.09, significant shrink-swell activity can be expected in such a soil.

Soil Moisture Parameters

Inferences that can be made from a knowledge of soil moisture data include the following:

- As water retention at 1500kPa has been equated with a $0.2\mu m$ pore diameter, a high correlation exists between it and clay content. For example, a good rule of thumb is that for soils dominated by silicates that disperse well in the standard particle size analysis,

$$\frac{\text{Water Held at 1500kPa}}{\% \, Clay} = 0.4$$

This could also be interpreted as *% Clay = 2.5 x Water Held at 1500kPa.* This could be a convenient way of cross-checking on the accuracy of the clay content of a soil from particle size distribution analysis.

- It is also known that a number of soil related factors can cause a deviation from the 0.4 reference point and can further aid in soil mineralogical data estimation. For instance, low activity clays such as kaolinites, chlorites, and some micas, tend to lower the ratio to less than or equal to 0.35. High activity clays such as smectites and some vermiculites, tend to increase the ratio as follows: 0.45 for smectites; 0.42 for micas and 0.32 for the Bt horizons of Paleudults (Soil Survey Staff, 1995).

- Organic matter content of a soil also tends to increase the ratio because it increases the water retention at 15-bar or 1500kPa. According to the US National Soil Survey Laboratory Staff (1983), an increased 15-bar to clay ratio can be expected if the organic C percent is >0.1 of the percent clay.

- If the ratio is ever greater than 0.6, and soil related factors do not adequately explain the situation, incomplete dispersion in particle size analysis may be a factor. Soil components which act as cements and cause poor dispersion include carbonates, gypsum, Fe oxides, and poorly crystalline Si.

- Clay-size carbonates tend to decrease the ratio in most cases. The 15-bar water retention for carbonate clays is approximately 2/3 the corresponding value for non-carbonate clays (Nettleton et al., 1991).

- Complete soil moisture characteristic curves are often needed to enable the relating of the measurements to the "moist" and "dry" definitions of the Xeric, Ustic, Udic, and Aridic soil moisture regimes of Soil Taxonomy. Saturation, a requirement of the aquic soil moisture regime, is also usually determined by recording the water table depth every few weeks throughout the year.

Soil pH

A number of inferences can be drawn from pH measurements in mineral soils. For instance, the following inferences regarding pH measurements can be made;

- A pH <3.5 associated with significant pH drop after wetting-drying cycles is an indication of the presence of acid sulphates often associated with Acid Sulphate Soils especially if the soils are from a coastal marsh enriched with soluble salts (ocean water). If the soil sample is however from mine spoil or from pits, it may suggest that buried sulphides in geologic materials have been oxidized (Buol et al., 1997).

- A value of pH 4.5-5.5 may be an indication that significant amounts of exchangeable Al and exchangeable H are present to affect plant growth and the base saturation percentage of the soil is low (Kamprath, 1967).

- Soluble bicarbonates seldom exceed 3 or 4 meq L^{-1} if pH is less than or equal to 7 (U.S Salinity Laboratory Staff, 1954).

- Soil pH 8.0 – 8.5 is an indication that the soil is fully base saturated and free $CaCO_3$ is present in the system and the exchangeable cation population is largely Ca + Mg. Also, soluble Ca^{2+} is seldom >2 meq L^{-1} if the pH >9.

- Soil pH 8.5 – 10 is an indication that the soil contains large amounts of soluble salts, especially exchangeable Na and the conductivity is mostly high.

- Gypsiferous soils seldom have pH >8.2.

- Measurements of pH in 1$\underline{N}$ KCl(pHs) are often useful particularly if compared with the pH in water (pHw). If the pHs is one-half or more units less than pHw in the range below pH 5.5, then significant amount of exch. Al or complexed slowly exch. Al are present as the drop in pH is due to the hydrolysis of the Al displaced by the K. If on the other hand, the pHs is greater than pHw, then the soil has a net positive charge due to the dominance of the exchange complex by hydrous iron oxides which is a good indication of the "acric" Great groups in the Oxisol Order. The increase in pH is due to the displacement of OH^s by the Cl^-.

- The NaF pH test is used as a simple and convenient index of andic materials. As a rule of thumb, a NaF pH > 9.4 is used as a strong indicator that andic materials (allophone) dominate the soil exchange complex (Fieldes and Perrott, 1966). The following reactions illustrate the action and form the basis of the procedure:

$$Al(OH)_3 + 3F^- \longrightarrow AlF_3 + 3 OH^-$$
$$Si(OH)_4 + 4F^- \longrightarrow SiF_4 + 4 OH^-$$

The amorphous material in the soil controls the release of the OH$^-$ and the subsequent increase in pH.

Organic Carbon/Total Nitrogen

Inferences derived from organic carbon data in soil classification activity are as follows:

- The content of organic carbon is used as the basis for differentiating between organic soils (Histosols) and mineral soils. In this regard, organic soils are defined as soils mainly composed of organic soil materials, while organic soil materials are defined as accumulated organic substances which if saturated with water for prolonged periods or if artificially drained, have more than 18% organic C if the mineral fraction has 60% or more clay or more than 12% organic carbon if no clay is present; or proportionately more organic carbon than a line connecting the two points for intervening clay contents (0 – 60%) e.g >15% organic C if mineral fraction contains 30% or more clay.

- All the eight epipedons recognized in Soil Taxonomy are principally differentiated on the basis of differences in organic carbon content or its components. For example, a histic epipedon is a surface horizon that contains more than 12 to 18% organic carbon, while a melanic epipedon contains 6% or more organic C but is at least 30cm thick. Also Mollic and Umbric epipedons must contain at least 0.6% organic C when mixed to a depth of 18cm while an Ochric epipedon contains less than 0.6% or fails to meet the depth requirements of the mollic or other epipedons amongst other requirements.

- Profile- depth distribution of organic carbon is a major differentiating characteristic in certain suborders of Soil Taxonomy. For example, one of the major criterion for the identification of Fluvents or Fluventic subgroups is the content of either 0.2% or more organic C of Holocene age at a depth of 125cm below the mineral soil surface or an irregular decrease in content of organic C from a depth of 25cm to a depth of 125cm or lithic or paralithic contact.

- The ratio of organic C to total N, C/N, is a useful indication of the degree of decomposition of organic matter in soils, as it narrows with increasing modification. Well decomposed soil humus has a C/N ratio of 12 to 13 in humid temperate soils, whereas straw, saw dust, lignin materials, etc., have much higher values (Buol et al., 1997). According to Young & Aldag (1982), the level of "fixed" N in many soils usually remains constant or increases with depth while organic C usually diminishes with depth resulting in a C:N ratio which narrows with depth. The potential to "fix" N has important fertility implications as the "fixed" N is slowly available for plant growth. It is also pertinent to note that uncultivated soils usually have higher C:N ratios than do cultivated soils. The C:N ratio of cultivated mollic epipedons have been known to vary from about 10 to 12, while forest soils have been found to be a few units higher (Young & Aldag, 1982).

Electrical Conductivity (EC)

The following inferences are made from data related to the electrical conductivity of the saturation extract in soil classification activity:

- A conductivity of saturation extract of 2 dSm^{-1} (25^0C) or 0.2 Sm^{-1} or greater in the upper part of the profile is used as one of the diagnostic criteria for the recognition of the Order Aridisols.

- A conductivity >4 dSm^{-1} (25^0C) is used to define a saline soil (U.S. Salinity Laboratory Staff, 1995).

- One of the major criterion for recognizing a Salic diagnostic horizon is that it must have an EC > 30 dSm^{-1} more than 90 days of the year. Also, a product of the EC in dSm^{-1} and thickness of the horizon in cm, must be equal to or greater than 900. A salic horizon within the upper 100cm of the soil surface is diagnostic for the Suborder Salids in the Aridisol Order.

Extractable (Free) Iron

Extractable or Free Iron refers to the total iron in a soil occurring as reductant-soluble hydrous oxides which are uncombined with layer silicate structures and are either in the form of clay particle coatings, or as discrete particles or possibly in interlayer positions. Free iron oxides contribute to greater soil aggregate stability and are active in phosphate fixation. Inferences related to soil classification from free iron in soils are as follows:

- In soils with weatherable iron-bearing minerals, the percentage of free iron increases with increasing weathering and soil age.

- Free iron decreases with increasingly poor drainage.

- Dithionite-citrate extractable iron is used as a criterion in the ferritic and oxidic mineralogy classes at the Family level of the USDA Soil Taxonomy and is used for setting limits for the ferritic, sesquic, ferruginous, etc classes (Soil Survey Staff, 1975, 1994).

- The content of extractable Fe is used together with extractable carbon and aluminium to define and recognize Spodic diagnostic horizons with their illuvial accumulation of Fe and/or Al and humus (Soil Survey Staff, 1999).

Cation Exchange Capacity (CEC)

CEC is a measure of the capacity of a soil to sorb or hold cations and to exchange species of these cations in reversible chemical reactions. This is a quality that is important for both soil fertility-nutrition studies and soil genesis. Despite variations in methodologies for their determination and the results obtained, CEC values are very useful in evaluating the capacity of soils to retain cations, their degree of weathering and general chemical reactivity (Buol et al., 1997).

Inferences often derived from CEC in pedological studies are as follows:

- The CEC determined by NH_4OAc at pH 7.0 often termed as CEC-7 to clay ratio is now used as auxiliary data to assess clay mineralogy. The ratio is an index for clay activity i.e., probable contribution of clay to the exchange capacity and soil solution chemistry. The smectites and vermiculites are considered high activity clays; kaolinites and hydroxyl-interlayered vermiculites are low-activity clays; while illites and chlorites are intermediate-activity clays (Soil Survey Laboratory Staff, 1983). The following guidelines for determining taxonomic soil mineralogy at the family level of Soil Taxonomy have been developed for the suite of montmorillonitic-mixed-kaolinitic materials:

CEC-7/Clay Ratio	Family Mineralogical Class
>0.7	Montmorillonitic
0.5-0.7	Montmorillonitic or Mixed
0.3-0.5	Mixed or Illitic
0.2-0.3	Kaolinitic or Mixed
< 0.2	Kaolinitic

The CEC-7/ Clay ratio thus serves as both an internal check of the data and as an estimator of clay mineralogy when mineralogy data are not available.

- When CEC is measured by NH_4OAc at pH 7, the values have been used by Rhoades (1982) to estimate the clay mineralogy in cmol kg^{-1} clay as follows:

CEC (in cmol kg^{-1} clay)	Clay Mineral Species
0	Sesquioxides
5-10	Halloysite $2H_2O$
2-16	Kaolinite
10-40	Chlorite
20-40	Illite (Clay mica)
40-50	Halloysite $4H_2O$
60-100	Montmorillonite
100-150	Vermiculite
160 (at pH 8.2)	"Amorphous" Clay
200-400	Organic matter

- As a rule, CEC-8.2 > CEC-7 > ECEC and the CEC-8.2 minus CEC-7 is used as an estimate of the pH dependent (variable) charge in a soil from pH 7.0 to 8.2.

- Kandic and Oxic Subgroups of Soil Taxonomy are differentiated on the basis of their low to very low CEC per kg clay; often < 16 cmol kg^{-1} at pH 7 or < 12 cmol kg^{-1} clay ECEC in their kandic, argillic or oxic B horizons.

- CEC data are used in the calculation of the Percentage Base Saturation (PBS) which is a widely used index for pedological and nutritional evaluation of soil quality.

- CEC at pH 7.0 is used in the calculation of Exchangeable Sodium Percentage (ESP) = $\frac{Exch. Na^+}{CEC (pH 7)}$ x 100 which is used in turn for the assessment of the sodicity of a soil. In this regard, a sodic soil among other properties is a soil which has an ESP greater than or equal to 15%. A Natric diagnostic subsurface horizon is defined as "a horizon that meets the requirements of an argillic horizon but also has prismatic or columnar structure, and over 15% of the CEC is saturated with Na^+ or has more exchangeable Mg^{++} plus Na^+ than Ca^{++} plus exchange acidity at pH 8.2" (Buol et al., 1997).

Percentage Base Saturation (PBS)

As earlier discussed, PBS is used in making of the following inferences:

- A major use of PBS is in the distinction between two Soil Orders; Alfisols and Ultisols. While Alfisols are expected to have PBS > 35% (by CEC determined by sum of cations at pH 8.2) or PBS > 50% (by CEC determined at pH 7.0) in the Bt horizon, Ultisols are required to have PBS <35 or <50% respectively in the Bt horizon.

- The major distinction between a Mollic and an Umbric epipedon is that mollic epipedon must have a PBS > 50% (by CEC at pH 7) while umbric epipedon has less than 50% by the same method of determination. Thus, since a mollic epipedon is diagnostic for the Mollisol Soil Order, the PBS is a major differentiate for three of the soil orders of the USDA Soil Taxonomy; Alfisols, Ultisols, and Mollisols as well as many soil great groups.

Extractable P and Phosphate Retention

The following inferences can be drawn from data related to extractable P and Phosphate retention:

- A Mollic and an Anthropic epipedons are separated on the basis that an Anthropic epipedon has equal to or greater than 1,500 mg kg^{-1} of citrate acid extractable P_2O_5 while a Mollic epipedon contains less than 1,500 mg kg^{-1} $P_2O_{5(acid\text{-}soluble\ phosphate)}$.

- Phosphate retention of 85% or more is one of the requirements for the identification of Andic soil properties which is diagnostic for the recognition of the Andisol Soil Order of Soil Taxonomy.

- P retention determination helps in the identification of soils in which P-fixation may be a problem affecting agronomic uses.

Mineralogical Properties

Inferences from mineralogical data can essentially be divided into two broad categories (a) Skeletal or Fine Sand Fraction Mineralogy and (b) Clay Mineralogy.

The following inferences can be made regarding the sand-size mineralogy of soil:

- Knowledge of the nature and conditions of the minerals in the fine sand fraction provides information on the source of parent materials; on the presence of lithological discontinuities or overlays in the solum or between the solum and the underlying material; and on the degree of weathering in the soil as a key to its history, genetic processes and possible fertility reserve (Cady et al, 1986). For instance (1) presence of lithologic discontinuities can be detected based on shifts in mineral species percentages from one horizon to the other (2) the higher the degree of weathering, the lower will be the ratio of weatherable to resistant minerals (3) the types of primary minerals present in a soil suggests the probable course of clay mineral formation and soil development (4) the higher the content of weatherable minerals in a soil which are capable of releasing plant nutrients as they weather, the higher the soil fertility reserve.

- Sand-size mineralogy data are also used as taxonomic criteria to classify pedons in soil mineralogy families of Soil Taxonomy (Soil Survey Staff, 2003).

Inferences related to clay mineralogy include the following:

- Horizons within the control section of a pedon are weighted to derive the family mineralogy placement of a soil body using clay mineralogical data.
- Soil physical qualities such as shrink-swell potential, plasticity, moisture retention and permeability are best estimated from clay mineralogical data.
- As earlier discussed, cation exchange characteristics are better determined with a knowledge of the clay mineralogical data of a soil.
- The weathering potential and the stage of weathering in a profile are better understood with clay mineralogical data.

Soil Micromorphology

Micromorphology may be defined as the study of micro-fabrics of soils in their natural undisturbed arrangement (Cady, 1965). Examination of thin sections (natural fabric of fine-earth fraction) with a polarizing light microscope can be considered an extension of field morphological studies. The results of micromorphological studies are most useful when they are combined with other field and laboratory information. Micromorphology is used to identify illuviation argillans, fabric types, skeleton grains, and weathering intensity as well as to investigate genesis of soil or pedological features. Soil Micromorphology is fast becoming a sub-discipline of Pedology but is beyond the scope of this book.

Revision Questions

1. (a) Describe the laboratory methods for the determination of the following soil properties often used in pedological studies.
   (i) Bulk density
   (ii) Coefficient of Linear Extensibility (COLE)
   (iii) Organic carbon
   (iv) Soil pH
   (v) Electrical conductivity of saturation extract

   (b) State two inferences relevant to soil classification that can possibly be drawn from each of the properties in Question 1a.

2. (a) List the three methods often used in the determination of Cation exchange Capacity (CEC) in soil classification activity.
   (c) Explain the usefulness of the results obtained by each of the methods in the elucidation of other pedological parameters used in soil classification activity.

3. Discuss the use of extractable (free) iron, extractable phosphorus and phosphate retention data in soil classification activity.

4. Discuss the use of mineralogical data in soil classification activity.

References

Blake, G. R. 1965. Bulk density. In C.A. Black (ed). Methods of Soil analysis. Agron. 9 Am. Soc. Agron., Madison, WI. Pp 374-390

Bray, R. H. and L. T. Kurtz, 1945. Determination of total, organic and available forms of phosphorus in soils. Soil Sci: 59:39-45.

Bremner, J. M. 1965. Total nitrogen. In C.A. Black (ed). Methods of Soil analysis, Agron. 9. Am. Soc. Agron., Madison. WI pp. 1149-1178.

Buol, S. W., F. D. Hole; R. J. McCracken and R. J. Southard. 1997. Soil Genesis and Classification, 4th ed. Iowa State University Press, Ames. 527p.

Cady, J. G., L. P. Wilding, and L.R. Dress. 1986. Petrographic microscope techniques. In A. Klute (ed.) Methods of soil analysis. Part I. Physical and Mineralogical properties. 2nd ed. Agron. 9: 185-218.

Day, P. R. 1965. Particle fractionation and particle-size analysis. In C.A. Black (ed). Methods of Soil analysis. Agron. 9. Am. Soc. Agron., Madison. WI. Pp. 545-567.

Fieldes, M. and K. W. Perrott. 1966. The nature of allophone in soils. Part 3. Rapid field and laboratory test for allophone. New Zealand Journ. Sci. 9: 623-629.

Grossman, R. B., B. R. Brasher, D. P. Franzmeier and J. L. Walker, 1968. Linear extensibility as calculated from natural-clod bulk density measurements. Soil Sci. Soc. Am Proc. 32:570-573.

Kamprath, E. J. 1967. Soil acidity and response to liming. Tech. Bull, Int. Soil Testing Service. Soil Sci. Dept., NC State Univ. Raleigh.

Lynn, W.C., W. E. Mckenzie, and R. B. Grossman. 1974. Field laboratory tests for characterization of Histosols. In Histosols. SSSA Spec. Publ. No. 6. Soil Sci. Soc. Am., Madison, Wisconsin.

Mehra, O. P. and M. L. Jackson. 1960. Iron oxide removal from soils and clays by a dithionite-citrate system with sodium bicarbonate buffer. Clays Clay Miner. 7:317-327.

Nettleton, W. D., B. R. Brasher and S. L. Banid. 1991. Carbonate clay characterization by Statistical methods. Soil Sci. Soc. Am. Special Publ. No. 26: 75-88.

Olsen, S. R. and L. E. Sommers. 1982. Phosphorus. In. A. L. Page *et al* (eds). Methods of soil analysis. Part 2. Chemical and microbiological properties. 2nd ed. Agron. 9: 403-430.

Rhoades, J. D. 1982. Cation exchange capacity. In A.L. Page *et al* (eds) Methods of Soil analysis. Part 2 Chemical and microbiological properties. 2nd ed. Agron 9: 149-157.

Richards, L. A. 1965. Physical condition of water in soil. In C.A. Black, ed. Methods of soil analysis. Agron. 9. Am. Soc. Agron. Madison, WI. Pp. 128-152.

SCS-USDA, 1972. Land resource regions and major land resource areas. M.E. Austin: US Dept. Agric. Handb. 296. U.S Govt. Printing Office, Washington, DC.

Soil Survey Laboratory Staff. 1983. National Soils Handb. SCS-USDA. U.S. Govt. Printing Office, Washington D.C.

Soil Survey Staff. 1975. Soil Taxonomy. USDA Handb. 436. US Govt. Printing Office, Washington D.C

Soil Survey Staff. 1994. Keys to Soil Taxonomy. USDA-SCS. 6th ed. Washington, DC.

Soil Survey Staff. 1995. Soil Survey Laboratory Information Manual. Soil survey Investigations Report No. 45 Version 1.0. 305p.

Soil Survey Staff. 1999. Soil Taxonomy. USDA Handb. 436 (2nd ed.) US. Govt. Printing Office, Wash. DC.

Soil Survey Staff. 2003. Keys to Soil Taxonomy. USDA-NRCS. 9th ed. Wash. DC. 332p.

U.S. Salinity Laboratory Staff. 1954. L.A. Richards (ed.). Diagnosis and improvement of saline and alkali soils. USDA Handb. 60. U.S Govt. Printing Office, Wash. DC.

Young, J. L. and R. W. Aldag. 1982. Inorganic forms of nitrogen in soil. In. J. F. Stevenson (ed.) Nitrogen in agricultural soils. Agron. 22: 43-66.

9

Principles of Soil Classification

General Concepts of Classification

Classification (also called Taxonomy) involves the grouping of objects, in te mind, on the basis of one or more properties. The objects to be classified are commonly called individuals. The individual is the smallest natural body that can be defined as a thing in itself. All the individuals of a natural phenomenon, collectively, are a population or a universe. Plants, animals, rocks, atoms and soils for example, are populations, each consisting of many individuals.

A class also called taxon (pl. taxa), is a group of individuals similar in one or more selected properties and distinguished from all other classes of the same population by differences in these properties. These properties, selected in accordance with the purpose of the classification, are termed differentiating characteristics or differentiae. They serve to differentiate among classes, i.e. to distinguish one class from all others.

When classifying the individuals of a large and widely varying population, such as plants and soils, classification is necessary not only of individuals into classes, but also of classes into wider or higher classes, and of those into still higher classes. This is called a **Multiple Category** as well as a **Hierarchical system** of classification. The individuals are grouped into classes of the lowest category, which are subsequently grouped into classes of higher categories. The highest categories have small numbers of classes defined in broad general terms, by means of a few differentiating characteristics. In the lower categories, there are large numbers of classes of narrow range defined in quite specific terms by a large number of differentiating characteristics.

Definition of Soil Classification

Judging from the foregoing discussion, Soil Classification may be defined as the systematic arrangement of soils into groups or categories on the basis of their characteristics. Broad groupings are made on the basis of general characteristics and subdivisions on the basis of more detailed differences in specific properties.

Early systems of soil classification were quite simple and highly practical. However, with increasing sophistication of agriculture, greater knowledge about soils as a collection of independent natural bodies, and greater complexity and diversity of soil uses, the classification of soils has become more scientific and organized.

Purposes of Soil Classification

From a practical point of view, classification enables the human mind to remember the properties of the objects being classified and make predictions about the behaviour of the objects as well as identify their best use. Classification also provides the basis for a common "language" by means of with scientists can exchange ideas and provide information for practical purposes.

In specific terms, the purposes of soil classification may be enumerated as follows:

1. To organize knowledge related to soils of the world, a region or a locality.
2. To understand relationships among soil individuals being classified and possibly learn new ones.
3. To more easily remember properties of the soils being classified.
4. To establish groups of the soils being classified for the following practical purposes:
    (a) predicting behaviour
    (b) identifying prime or unique agricultural lands suited to specific land uses.
    (c) Estimating the productivity of soils.
    (d) Extrapolation or transfer of knowledge of one soil to other areas by providing a common "language" by means of which scientists can exchange ideas and provide information for the solution of practical problems (a soil map).

Steps Involved in Soil Classification

The procedure of carrying out soil classification involves a number of steps which ultimately depends on the final objective of the classification but most often involves the following steps:

(a)   Comprehensive study of the physical environment in which the soils to be classified are located. The parameters so studied include the geology, geomorphology, vegetation, landuse, drainage, and climate of the area.

(b)   Field mapping, soil characterization and field sampling of soils is then carried out.

(c)   More adequate knowledge of the soil characteristics is then sort through morphological and micromorphological study of the soils; laboratory characterization of the fine earth fraction of each horizon within pedons; elucidation of genetic factors of soil formation and the relation between the soils and the environment in which they are located.

(d)   Armed with an adequate knowledge of the characteristics of the soils, the soil classifier (pedologist) then formulates differentiae at the lowest category.

(e)   The next step involves the clustering of soil individuals at lower categories into higher level taxa or classes on the basis of similar characteristics (Multiple Category System).

(f)   This is followed by the regrouping of higher classes into yet higher class (Hierarchical System)

(g)   More extensive soil mapping is then carried out in areas of similar environmental setting to test the classification carried out at a different location.

(h)   The classification is then tested and adapted several times so as to discover and learn new relationships. At this point, the classification is sent out to colleagues for necessary inputs and valid criticisms.

(i)   The new relationships learnt, the criticisms and other ideas are finally incorporated into the classification and a Final Classification is thus obtained.

Kinds of Soil Classification

In general, two broad kinds of soil classification are recognized; a Natural or Taxonomic Classification also termed Scientific Classification and a Technical Classification.

A Natural Classification is one in which the purpose of the classification is, in so far as possible, to bring out relationships of the most important properties of the population being classified without reference to any single specified and applied objective. In a natural classification, all the attributes of a population are considered and those which have the greatest number of covariant or associated characteristics are selected as the ones to define and separate the various classes. Most soil classification systems try to approach a natural classification system as an ideal, though some more weight tend to be given to properties of higher agricultural relevance. The most common examples of a natural soil classification system are the USDA Soil Taxonomy System (Soil Survey Staff, 1975, 1999) and the FAO-UNESCO Soil Classification System (FAO-UNESCO, 1988) or presently the World Reference Base for Soil Resources (FAO, 1998). These two systems have received world wide attention and are the most used in Nigeria. The salient features of these systems of soil classification will be briefly discussed later.

A Technical Classification, on the other hand, is one which is aimed at a specific, applied, practical purpose. For example, classifying soils for agriculture or engineering purposes or even more specifically classifying soils for maize production or for irrigated agriculture. The most common examples of a technical classification which are widely used world wide are the USDA Land Capability Classification System and the FAO Land Suitability Classification for rainfed agriculture. These two examples are discussed along with others in chapter 11 of this book.

USDA Soil Taxonomy System

In 1951, the Soil Survey Staff, Soil Conservation Service of the United States Department of Agriculture (USDA) started the development of a new system of soil classification. The system was developed by a series of approximations, testing each one to discover its defects and thus gradually approaching a workable system. In 1960, the *7th Approximation* was published to secure the widest possible criticism. It was adapted and changed in various supplements and in 1975, the final text was published under the title "**Soil Taxonomy: A Basic System of Soil Classification for Making and Interpreting Soil Surveys**". In the 1975 edition, ten Soil Orders were recognized (Soil Survey Staff, 1975), but several revisions have since been carried out such that in 1994, one additional Soil Order of **Andisols** was added making it a total of eleven

Soil Orders (Soil Survey Staff, 1994). In 1998, further revision of the Taxonomy led to the creation of a further Soil Order, **Gelisols**, to accommodate soils with gelic materials underlain by permafrost with freezing and thawing processes common in very cold regions of the world. This has brought the total number of Orders in the Taxonomy to twelve (12) at present (Soil Survey Staff, 1999).

The System is a Multiple Category as well as Hierarchical System and contains six categories. From highest to lowest levels of generalization, they include; the Order, Suborder, Great group, Subgroup, Family and Series. Figure 9.1 contains a listing of the hierarchy of the categories and the number of taxonomic classes (taxa) they each contain.

The system is said to be **Hierarchical** because the lower categories fit within the higher categories. Thus, each Order has many Suborders, each Suborder has several Great groups and each Great group even more or several subgroups, etc. as shown in Fig. 9.1

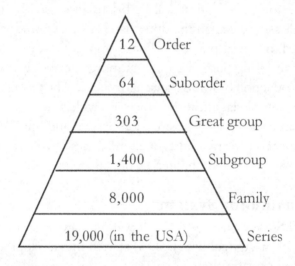

Fig. 9.1: *Multiple Category and Hierarchical System of the USDA Soil Taxonomy*

Two main features which make the USDA Soil Taxonomy System unique are the fact that (i) The system is based on soil properties that can be objectively observed in the field or measured in the laboratory; so it is not based on presumed mechanisms of soil formation (ii) it employs a unique nomenclature which gives a definite connotation of the major characteristics of the soils in questions.

A number of differentiating characteristics or differentiae are used in the separation of the various categories of taxa of the Soil Taxonomy, such as Diagnostic Surface Horizons (Epipedons) and Diagnostic Subsurface Horizons (Endopedons) for Mineral Soils; Diagnostic Characteristics for Organic Soils; and Other Diagnostic Features for Mineral and Organic Soils as well as Soil Moisture and Soil Temperature Regimes. We shall proceed to define these differentiae as follows:

Diagnostic Surface Horizons or Epipedons

Epipedons are horizons that are formed at or near the surface and in which most of the rock structure has been destroyed. It is darkened by organic matter or shows evidence of eluviation or both (USDA-NCRS, 2003). An Epipedon is not synonymous with the A Horizon always because it may include part or all of the E, AB or BA or even part or all of the illuvial B Horizon, if the darkening by organic matter extends from the soil surface into or through the B Horizon. There are **eight** epipedons presently recognized in the Soil Taxonomy defined as follows:

Mollic Epipedon is a mineral surface horizon >25cm thick, noted for its dark colour, high organic carbon content (>0.6% throughout its thickness) and softness even when dry. It has a high percentage base saturation (>50% by NH_4OAc at pH 7.0). Mollic epipedons form from the underground decomposition of mainly grass organic residues and are associated with Mollisols of the grassland or prairie regions of the world. They have good soil structure and porosity and are associated with some of the richest agricultural soils of the world.

Umbric Epipedon is a surface horizon that is similar to a mollic epipedon in several respects except that they are formed largely under forest vegetation and have a lower natural fertility status than a mollic epipedon. Thus, the major distinction between them is that an umbric epipedon has a percentage base saturation of <50% by NH_4OAc at pH 7. It is most commonly associated with Inceptisols, Andisols, Alfisols, and Ultisols.

Anthropic Epipedon is a surface horizon formed from long-term continuous use by humans and consequent high return of organic residues to the surrounding environment. Thus, anthropic epipedons have the dark colour of mollic or umbric epipedons in addition to containing over 1500 $mgkg^{-1}$ of 1% citric acid soluble phosphate (P_2O_5) and a regular decrease in P_2O_5 to a depth of 125cm. They occur in a variety of soil orders.

Plaggen Epipedon is a human-made surface layer >50cm thick that has been produced by long-continued manuring. They occur mostly in Europe where in medieval times, sod or other materials were used for bedding livestock and the manure was spread on cultivated fields. They often contain artifacts, such as bits of brick and pottery throughout its depth. The plaggen epipedon is associated with Inceptisols.

Folistic Epipedon is a surface horizon which is 20cm or more thick and contains high amounts of organic C with bulk density less than 0.1 gm cm^{-3}, but is not saturated with water for more than a few days after heavy rains. They occur mostly in cool, humid regions under forest vegetation and are associated mostly with Spodosols and Inceptisols.

Histic Epipedon is a layer that is 20 to 60 cm thick and contains 75% or more (by volume) fibres with a bulk density of <0.1gm cm^{-3} just as in the folistic epipedon, but differs from the latter in that it is commonly saturated with water for 30 days or more, cumulative during normal years, unless artificially drained. They can occur in many soil orders.

Melanic Epipedon is a thick, black horizon that contains high concentrations of organic C and short-range order minerals (allophane, imogolite, or ferrihydrite) or Al-humus complexes. It occurs in areas associated with volcanic influence under forest vegetation or the soil order Andisols.

Ochric Epipedon is a surface horizon which fails to meet the criteria of any of the other seven diagnostic epipedons: For instance, it is either too light in colour, too low in organic C and if high in organic C or dark in colour, it is too thin to be a mollic epipedon. Ochric epipedon can occur in a variety of soil orders but is most common in the tropics.

Diagnostic Subsurface Horizons or Endopedons

Endopedons or the Subsurface Horizons form below the soil surface mainly through pedogenetic processes, but can be exposed at or near the surface in eroded soils. They are **nineteen** in number and are regarded mostly as the B horizon by many Pedologists. They are briefly described as follows:

Agric Horizon is an illuvial horizon >10cm thick that has formed under cultivation and is directly below an Ap horizon and contains significant amounts of illuvial silt, clay and humus. They often contain 5% or more (by volume) worm holes or lamellae that have thickness of 5mm or more and have a value (moist) of 4 or less and chroma of 2 or less. They are often associated with Alfisols.

Albic Horizon is an eluvial horizon 10cm or more thick often with a bleached or white colour because clay and Fe oxides coatings have been removed by pedogenesis. They often overlie an argillic, kandic, natric, cambic or spodic horizon or a fragipan (all defined later) and are associated with Spodosols, Alfisols, Ultisols and Mollisols.

Argillic Horizon is commonly a subsurface horizon at least >15cm thick or 1/10 as thick as the overlying layer, with significantly higher content of illuvial clay than the overlying horizon. It must have at least 1.2 times as much clay as the eluvial horizon. For instance, if an eluvial horizon has <15% clay, then the argillic horizon must have at least 3% more clay or 8% more clay if the eluvial horizon has >40% clay. Argillic horizon must show evidence of clay illuviation such as clay cutans or argillans. They occur on stable landscapes and are mostly associated with Alfisols, Ultisols, Aridisols and Mollisols.

Calcic Horizon is an illuvial horizon >15cm thick in which secondary $CaCO_3$ has accumulated to a significant extent. It should contain 15% or more $CaCO_3$ equivalent and its $CaCO_3$ equivalent should be more than 5% or more higher than that of an underlying horizon. It should not be indurated or cemented. Calcic horizons occur in Aridisols, Mollisols, Inceptisols, and Gelisols.

Cambic Horizon is an altered horizon >15cm thick which results from physical alterations, chemical transformations, or removals or a combination of two or more of these processes. Its soil texture varies from very fine sand, loamy fine sand or finer with some weak indication of either an argillic or spodic horizon, but not enough to qualify as either, e.g. Less than 1.2 times as much clay as an overlying horizon. They are commonly identified by structure and a higher clay content, redder hue, or higher chroma than an overlying horizon. They are most commonly associated with Inceptisols, but they can also occur in Aridisols, Mollisols, Vertisols, Andisols, and Gelisols.

Duripan is an illuvial silica-cemented subsurface horizon with or without auxiliary cementing agents. The pan shows evidence of the accumulation of opal or other forms of silica such as coatings, lenses and bridges between sand-sized grains. Air-dry fragments from more than 50% of the horizon do not slake in water or HCl but do slake in hot Conc.KOH. Duripans limit the downward movement of roots and water. They are most commonly associated with Aridisols, but they can occur in Alfisols, Andisols, Inceptisols and Spodosols.

Fragipan is a non-cemented, brittle when moist and very hard when dry, subsoil layer >15cm thick, with high bulk density of 1.6 to $1.8g.cm^{-3}$ which restricts the entry of water and roots into the soil matrix. It does not soften on wetting but can be broken in the hands. Air-dry fragments slake when immersed in water. They are commonly associated with Ultisols, Alfisols, Spodosols and Inceptisols.

Glossic Horizon is 5 cm or more thick and develops as a result of the degradation of an argillic, kandic, or natric horizon from which clay and free iron oxides are removed. It generally consists of an eluvial part i.e., albic materials, which constitute 15 to 85% (by volume) of the glossic horizon and an illuvial part, i.e., remnants (pieces) of an argillic, kandic, or natric horizon. They commonly occur in Alfisols and Ultisols.

Gypsic Horizon is a subsoil horizon <15cm thick of secondary gypsum accumulation. It contains at least 5% more gypsum than underlying material. Most gypsic horizons occur in arid environments where the parent material are rich in gypsum. In soils with groundwater close to the surface, gypsum can accumulate by capillary rise, evaporation and transpiration. They occur most commonly in Aridisols, but can also occur in a few Inceptisols and Gelisols.

Kandic Horizon is a vertically continuous subsurface horizon >18cm thick or 5cm thick if the textural transition is abrupt, that underlies a coarser textured surface horizon. The horizon is diagnostically composed of low-activity clays with < or equal to 16 cmol kg^{-1} clay CEC at pH 7 and < or equal to 12 cmol kg^{-1} clay ECEC (Total Bases + Exch Al). Just like the argillic horizon, it has a clay content increase at its upper boundary of >1.2 times clay within a vertical distance of <15cm, however, clay skins or cutans may or may not be present. Kandic horizons occur in Alfisols, Ultisols and Oxisols.

Natric Horizon meets the requirements of an argillic horizon but in addition has an exchangeable sodium percentage >15% or has more exch. Mg^{++} plus Na$^+$ than Ca^{++} plus exchange acidity at pH 8.2. Natric horizons are more commonly associated with the Aridisols, but they also occur in the Alfisols, Ultisols and Mollisols.

Ortstein is a cemented horizon >25mm thick that consists of complexes of Al and organic matter with or without Fe (spodic materials) which limits root growth and the downward movement of water. Ortstein occurs in Spodosols.

Oxic Horizon is a subsurface horizon at least 30cm thick with sandy loam or finer texture, low CEC and low amounts of weatherable minerals. It has a high content of low-activity clays with apparent ECEC of 12 or less cmol kg^{-1} clay and 16 or less cmol kg^{-1} clay CEC at pH 7. Clay content increase at the upper boundary is more gradual than required by the Kandic horizon. It contains <10% weatherable minerals in the sand fraction and has <5% (by volume) rock structure. They occur mainly in Oxisols.

Petrocalcic Horizon is a cemented or indurated Calcic horizon >10cm thick, whose fragments do not break down in water but does so in acid. They are associated with Aridisols but also occur in Alfisols, Mollisols and Inceptisols.

Petrogypsic Horizon is a cemented or indurated Gypsic horizon >10cm thick, whose dry flakes do not slake in water. They are commonly associated with Aridisols.

Placic Horizon is thin (<25mm), dark coloured (dark reddish brown to black) horizon that is cemented by either Mn or Fe and Mn and organic matter and lies within 50cm of the soil surface and is wavy, involuted and slowly permeable. They are associated mainly with Spodosols and Inceptisols.

Salic Horizon is a horizon of accumulation of secondary soluble salt enrichment at least 15cm thick such that the EC of saturation extract is <30 dSm^{-1} in more than 90 days of the year. Also a product of the EC in dSm^{-1} and thickness in cm is >900. They are mainly associated with Aridisols.

Sombric Horizon is a free-draining horizon, not under an Albic horizon, has the darkness and base saturation status of an umbric epipedon and has formed by illuviation of humus but not of Al or Na. It may easily be mistaken for a buried A horizon. They are largely confined to the cool, moist soils of high plateaus or mountains of the tropics and occur in Inceptisols, Oxisols and Ultisols.

Spodic Horizon is an illuvial horizon that contains active amorphous materials composed of organic matter and Al with or without Fe. They are associated with Spodosols.

Sulphuric Horizon is a mineral or organic soil horizon that has a pH <3.5, is toxic to plants roots, and has yellow mottles of jarosite. They occur in wet Entisols, Histosols and Inceptisols.

Diagnostic Characteristics for Organic Soils

An **Organic Soil** is a soil in which more than half of the profile thickness is composed of organic soil materials. On the other hand, **Organic Soil Materials** are pieces of plant tissues or fibres which are saturated with water for prolonged periods unless artificially drained and contain >18% organic carbon (by weight) if the mineral fraction contains more than 60% clay or >12% organic carbon if the mineral fraction has no clay or between 12 and 18% carbon if the clay content of the mineral fraction is between 0 and 60%. Organic soil materials are also fibres which are never saturated with water for more than a few days and having more than 20% organic carbon. Histosols develop on these Organic soil materials. There are various kinds of organic soil materials which are diagnostic for the classification of Organic Soils and described as follows:

Fibric Soil Materials are organic soil materials that contain three-fourths or more (by volume) of undecomposed fibres after rubbing, excluding coarse fragments. They are also referred to as Peat.

Hemic Soil Materials are intermediates (50%) in their degree of decomposition of organic soil materials between the less decomposed fibric and the more decomposed sapric materials. Their morphological features give intermediate values for fibre content, bulk density, and water content. They are referred to as Mucky-Peat or Peaty-Muck.

Sapric Soil Materials are the most highly (75%) decomposed of the three kinds of organic soil materials. They have the smallest amount of fibre, the highest bulk density, and the lowest water content on a dry weight basis at saturation. They are commonly very dark grey to black in colour.

Humilluvic Material is illuvial humus which accumulates in the lower parts of some organic soils that are acid and have been drained and cultivated. It has very high solubility in sodium pyrophosphate and rewets very slowly after drying.

Limnic Materials include both organic and inorganic materials that were either deposited in water by precipitation or through the action of aquatic organisms such as algae or diatoms or derived from underwater and floating aquatic plants and subsequently modified by aquatic animals. They include coprogenous earth (sedimentary peat), diatomaceous earth (geologic, siliceous diatoms) and marl (soft unconsolidated $CaCO_3$).

Other Diagnostic Soil Characteristics or Diagnostic Features

In addition to the diagnostic epipedons, endopedons and organic soil materials already discussed, several other diagnostic features are recognized and used in the USDA Soil Taxonomy at a somewhat lower level of the System in the classification of both mineral and/or organic soils and are briefly described as follows:

Abrupt Textural Change is a considerable increase (usually >20%) in clay within a short distance (7.5cm) usually associated with argillic or kandic horizons.

Andic Soil Properties are materials which contain significant amounts of volcanic materials and short-range-order minerals notably, allophane, imogolite, and ferrihydrite or Al-humus complexes. They have bulk density (at 33 kPa) of <0.90 gcm-3; a phosphate retention of >85% and Al+½ Fe (by ammonium oxalate extraction) of >0.4%.

Aquic Conditions refer to soils that currently experience continuous or periodic saturation and reduction as indicated by redoximorphic features. There are three types of saturation recognized: **Episaturation**, when due to perched water table; **Endosaturation**, when due to high ground water table and **Anthraquic** saturation when due to controlled flooding such as that are used to grow rice or sugar cane.

Albic Materials are soil materials in which the clay and/or the oxides coatings have been removed or eluviated such that its colour is largely determined by the colour of the primary sand and silt particles which is usually white to grey or light coloured, like the materials in an E horizon.

Durinodes are weakly cemented to indurated nodules with a diameter >1cm. The cement is SiO_2, presumably opal and microcrystalline forms of silica. Air-dry peds do not slake in water or HCl but will be destroyed by hot KOH after acid washing to remove carbonates.

Lamellae A Lamella is an illuvial horizon less than 7.5 cm thick formed in unconsolidated regolith more than 50 cm thick. Each lamella contains an accumulation of oriented silicate clay on ped faces or bridging sand and silt grains or coarser fragments if present. Lamellae occur in a vertical series of two or more, and each lamella must have an overlying eluvial horizon.

n Value According to Pons and Zonneveld (1965), the *n* value characterizes the relation between the water in a soil under field conditions and its percentages of inorganic clay and humus. The *n* value is helpful in predicting whether a soil can be grazed by livestock or can support other loads and in predicting what degree of subsidence would occur after drainage. It is calculated by the following formula:

n = (A – 0.2R) / (L + 3H) where A = % water in the soil in field conditions (on a dry-wt basis); R = % silt + sand; L = % clay; and H = % organic matter (% org. carbon x 1.724). The greater the n value the greater the risk of soil failure, with the value of 0.7 serving as the critical limit.

Lithic Contact is the boundary between soil and a coherent underlying material, notably hard bedrock. Cracks that can be penetrated by roots are few and their horizontal spacing is >10 cm.

Paralithic Contact is the boundary between soil and relatively unaltered materials such as partially weathered rock or unconsolidated material also termed paralithic materials.

Petroferric Contact is a boundary between soil and a continuous layer of indurated materials in which Fe is an important cement and organic matter is either absent or present in trace amounts.

Permafrost is defined as a thermal condition in which a soil material remains below 0^0C (frozen) for two or more years in succession. The frozen layer has a variety of ice lenses, veins, segregated ice crystals and ice wedges.

Plinthite is an Fe-rich, humus poor mixture of clay with quartz and other minerals or diluents commonly occurring as dark red redox concentrations that usually form platy, polygonal, or reticulate patterns. They harden irreversibly to ironstone hardpan or to irregular aggregates on exposure to repeated wetting and drying, especially the heat from the sun.

Soil Moisture Regimes are defined in terms of the level of ground water and in terms of the seasonal presence or absence of water held at a tension of less than 1500 kPa in the soil moisture control section. For most soil orders, soil moisture regime is used to determine placement of a soil at the suborder level. They include the following classifications:

- **Aquic moisture regime** signifies a reducing soil moisture regime virtually free of dissolved O_2 from saturation with water. **Peraquic** is the term used to describe soils that are nearly always saturated all year round.
- **Aridic or Torric moisture regime** applies to soils which commonly occur in arid climates and are dry in all parts more than half the time during the growing season and moist in some or all parts for less than 90 consecutive days during the growing season. Aridic is

used in dry regions while the term Torric is used in dry and hot environments. In practical terms, soils with aridic or torric soil moisture, do not have enough moisture to grow and mature a crop without irrigation (Buol et al., 1997).

- **Udic moisture regime** applies to soils which occur in climates with well-distributed rainfall such that, the total annual precipitation is equal to or exceeds the total annual evapotranspiration. The term **Perudic** soil moisture is used when in most years, precipitation exceeds evapotranspiration every month of the year. In practical terms, soils with udic moisture regime have adequate moisture for growing crops any time the soil temperature is satisfactory for crop growth while soils with perudic soil moisture have excess rainfall such that harvesting and curing a crop is difficult.

- **Ustic moisture regime** is moister than aridic and drier than udic. The concept of ustic soil moisture is one of limited moisture but at least some moisture at a time when conditions are suitable for plant growth.

- **Xeric moisture regime** apply to soils which occur only in temperate areas with cool, moist winters and warm dry summers. Most crop growth under xeric soil moisture, takes place in the spring utilizing stored soil moisture from the winter period.

Soil Temperature Regimes are based on measured mean annual temperature of a soil at 50 cm depth or at a lithic, paralithic, or densic contact, whichever is shallowest and are categorized as follows:

- **Cryic soil temperature regime** which applies to soils which have a mean annual soil temperature between 0^0 to 8^0C with summer temperatures less than 15^0C.

- **Frigid soil temperature** applies to soils which have a mean annual soil temperature of $<8^0$C.

- **Mesic soil temperature** applies to mean soil temperatures greater than or equal to 8^0C but $<15^0$C.

- **Thermic soil temperature** applies to soils with mean soil temperatures greater than 15^0C but $<22^0$C.

- **Hyperthermic soil temperature** applies to soil with mean soil temperatures greater or equal to 22^0C.

When the difference between the mean summer and mean winter soil temperature is less than 6^0C, **iso** is added to the name. However, only

Isofrigid, Isomesic, Isothermic and Isohyperthermic soil temperature regimes are used. Soil temperature is most commonly used at the Family level of the Soil Taxonomy.

Categories of the USDA Soil Taxonomy

A category of Soil Taxonomy is a set of classes that is defined approximately at the same level of generalization or abstraction and includes all soils. As earlier mentioned, there are six categories or levels in Soil Taxonomy. In order of decreasing rank and increasing number of differentiae and classes, the categories are Order, Suborder, Great group, Subgroup, Family and Series.

Soil Order

This category is based on soil forming processes as indicated by the presence or absence of major diagnostic horizons. Each of the twelve Soil Orders includes soils whose properties suggest that they are not dissimilar in their genesis. They are thought to have been formed by the same general genetic processes. The names, formative element, derivation, pronunciation and major characteristics of the twelve Orders arranged as much as possible from the least developed Entisols (no soil profile development) to the one with the greatest soil development, Oxisols (deeply weathered, highly leached, sesquioxidic soils of the tropics) are presented in Table 8.1. Pictorial representations of modal profiles of each of the twelve orders are also presented in Appendix 1.

Table 9.1: *Names of the Soil Orders in the USDA Soil Taxonomy with their derivation, formative elements, pronunciation, and major characteristics.*

Order Name	Formative Element	Derivation	Pronunciation	Major Characteristics
Entisols	Ent	Nonsense symbol	Recent	Soils of recent origin with little or no evidence of the development of pedogenic horizons except ochric or anthropic epipedons. Usually either very sandy or very shallow to lithic or paralithic contact because of their location on very unstable landscapes. Constitute 16% of earth surface.

Order Name	Formative Element	Derivation	Pronunciation	Major Characteristics
Inceptisols	ept	L.*inceptum*, beginning	Inception	Embryonic soils with few diagnostic features, ochric or umbric epipedons and cambic horizon. Often found on fairly steep slopes, young geomorphic surfaces and on resistant parent materials. Constitute 17% of the global ice-free land area, the largest of all soil orders.
Gelisols	el	Gk. *gelid*, very cold	Jelly	Soils of very cold climates that contain permafrost within 2m of the soil surface, often with cryoturbation (frost churning). Limited geographically to high-altitude polar regions and localized areas at high mountain elevations. They are characterized by a diagnostic perennial coldness and not diagnostic horizons. Constitute 8.6% of the earth land area.
Andisols	and	Jap.*ando*, black soil	*And*esite	Soils that have formed in volcanic ash or other volcanic ejecta with properties dominated by short-range-order minerals such as allophane, imogolite, ferrihydrite and organo-metallic complexes. Until 1989, were Andepts and Andaquepts subgroups of Inceptisols. They have andic soil properties, melanic and histic epipedons. Constitute only 0.75% of the earth land area and are the least in extent of the Orders.
Histosols	ist	Gk. h*istos*, tissue	*Hist*ology	Soils composed of organic soil materials containing at least >20% organic matter by weight and are more than 40cm thick with bulk densities <0.3 gcm^{-3}. Often formed in wetlands and cold environments where restricted drainage and cold temperatures inhibit OM

Order Name	Formative Element	Derivation	Pronunciation	Major Characteristics
				decomposition and allow organic materials to accumulate over time. They are also referred to as Peat and occupy 1% of the world's land area.
Aridisols	Id	L. *aridus*,dry	*Arid*	Soils of arid regions (cold or warm deserts) with cryic or frigid, thermic or hyperthemic soil temperature regimes and aridic or torric soil moisture regimes. They may have subsurface diagnostic horizons in which clays, $CaCO_3$, silica, salts, and/or gypsum have accumulated. Constitute 11.8% of the earth's land area.
Vertisols	ert	L. *verto*, turn	In*vert*	Dark, swelling, shrinking and cracking, clay soils which go through periods of wetting and drying. Have high content (>30%) of smectitic clays to a depth of 1m or more. Despite its often dark colour, OM content ranges only from 1% to 6%. They commonly develop from limestone, basalt, or other Ca & Mg rich parent materials and are found mostly in subhumid to semiarid, warm regions under grassland vegetation. Constitute 2.5% of global land area.
Mollisols	oll	L. *mollis*, soft	Mo*llify*	Dark, soft soils of grassland ecosystems occurring mainly in the Prairies of the USA and the Steppes of Russia. They are characterized by Mollic epipedon, Argillic or Natric horizons and are of high base status. Occupy about 7% of the world's total soil area.

Order Name	Formative Element	Derivation	Pronunciation	Major Characteristics
Alfisols	alf	Nonsense symbol	Ped*alfer*	Soils that have an argillic, kandic or natric horizon and a moderate to high base saturation >35% (at pH 8.2) or >50% (at pH 7.0). They typically have an ochric epipedon but may have an umbric epipedon. May also have a fragipan, plinthite, a duripan or a petrocalcic horizon. Occur on relatively old landscapes and occupy 13% of the earth land area.
Ultisols	ult	L. *ultimus*, last	*Ulti*mate	Strongly leached, low base status (<35% at pH 8.2 or <50% at pH 7), acid forest soils with relatively low native fertility. Same as Alfisols in other diagnostic properties and occupy 8.1% of the earth's land area.
Spodosols	od	GK. *Spodos*, wood ash	*Pod*zol: odd	Acid, sandy, forest soils with eye-catching and photogenic Albic horizon overlying a reddish-brown Spodic horizon. Often occur under Coniferous forest in cool, moist climates. Occupy 3.5% of earth's land area.
Oxisols	ox	Fr. O*xide*, Oxide	O*xide*	Very highly weathered, highly leached and sesquioxide-rich soils found primarily in the intertropical regions of the World. Characterized by the presence of Oxic horizon, very few weatherable minerals, high P retention & low CEC. Occupy 8% of the earth's land area.

Suborder

Classes at the suborder level are separated within each order on the basis of soil properties that reflect major environmental controls on the current set of soils-forming processes. Most Suborders are mostly indicative of the soil moisture regime or to a lesser extent the soil temperature regime under which the soils are found. For example, soils formed under wet conditions (aquic soil moisture regime) are generally identified under separate suborders such as Aquents, Aquerts and Aquents. While Cryalfs, Cryands and Cryids represent very cold (cryic) soil temperature regimes of Alfisols, Andisols and Aridisols. Table 9.2 gives a listing of the suborders under each of the twelve Soil Orders of **Soil Taxonomy**. Table 9.3 also lists the formative elements in Names of the suborders and relate them to their soil characteristics. Thus, the Udults suborder, which are predominant in southern Nigeria, are moist Ultisols. It is pertinent to note that the ending of the suborder name identifies the Order in which the soils are found.

Great Group

Great groups, about 300 of them, are subdivisions of suborders, and are defined largely by the presence or absence of diagnostic horizons and their arrangement. Great group names are made up of prefixes of formative elements derived from names of diagnostic horizons attached to the names of suborders in which the Great group occurs. For example Ustalfs with a natric horizon (high in sodium) belong to the Natrustalfs great group.

Udults located on old land surfaces belong to the Paleudults great group. Table 9.4 contains a listing of the common formative elements which are often combined with the appropriate suborder names to give the Great group names. The connotations of the formative elements are also given.

Table 9.2:   *Soil Orders and their Suborders in Soil Taxonomy*

Order	Suborder	Order	Suborder	Order	Suborder
Alfisols	Aqualfs	Andisols	Aquands	Aridisols	Argids
	Cryalfs		Cryands		Calcids
	Udalfs		Torrands		Cambids
	Ustalfs		Udands		Cryids
	Xeralfs		Ustands		Durids
			Vitrands		Gypsids
			Xerands		Salids
Entisols	Aquents				
	Arents	Gelisols	Histels	Histosols	Fibrists
	Fluvents		Orthels		Folists
	Orthents		Turbels		Hemists
	Psamments				Saprists
			Albolls		
Inceptisols	Anthrepts	Mollisols	Aquolls	Oxisols	Aquox
	Aquepts		Cryolls		Perox
	Cryepts		Rendolls		Torrox
	Udepts		Udolls		Udox
	Ustepts		Ustolls		Ustox
	Xerepts		Xerolls		
Spodosols					
	Aquods	Ultisols	Aquults	Vertisols	Aquerts
	Cryods		Humults		Cryerts
	Humods		Udults		Uderts
	Orthods		Ustults		Usterts
			Xerults		Xerets

Table 9.3: *Formative Elements in Names of Suborders in Soil Taxonomy*

Formative Element	Derivation	Connotation
alb	xer L.*albus,white*	Presence of albic horizon (a bleached eluvial horizon)
anthr	Gk. Anthropos, *human*	Presence of anthropic or plaggen epipedon
aqu	L. *aqua, water*	Characteristics associated with wetness
ar		Mixed horizons
arg	L. *arare*, to plow L. *argilla*, white clay	Presence of argillic horizon (a horizon with illuvial clay)
calc		Presence of calcic horizon
camb	L. *calcis*, lime	Presence of cambic horizon
cry	L. *cambriare*, to change	Cold
dur	GK. *Kryos*, icy cold	Presence of a duripan
fibr	L. *durus*, hard	Least decomposed stage
fluv	L. *fibra*, fiber	Floodplains
fol	L. *fluvius*, river	Mass of leaves
gyps	L. *folia*, leaf	Presence of gypsic horizon
hem	L. *gypsum*, gypsum GK. *Hemi*, half	Intermediate stage of decomposition
hist		Presence of histic epipedon
hum	GK. *Histos*, tissue	Presence of organic matter
orth	L. *humus*, earth	The common ones
per	GK. *Orthos*, true L. *per*, throughout time	Of year-round humid climates, perudic moisture regime
psamm		Sand textures
rend	GK. *Psammos*, sand	Rendzinalike – high in carbonates
sal	Modified from Rendzina	Presence of salic (saline) horizon
sapr	L. *sal*, salt	Most decomposed stage
torr	GK. *Sapros*, rotten	Usually dry
turb	L. *torridus*, hot and dry	Cryoturbation
ud	L. *turbidus*, disturbed	Of humid climate
ust	L. *udus*, humid L. *ustus*, burnt	Of dry climates, usually hot in summer
vitr		Resembling glass
	L. *vitreus*, glass GK. *Xeros, dry*	Dry summers, moist winters

Note: *L = Latin; GK = Greek Derivations*

Table 9.4: *Formative Elements for Names of Great Groups and their Connotation*

Formative	Connotation	Formative	Connotation	Formative	Connotation
acr	Extreme weathering	fol	Mass of leaves	petr	Cemented horizon
agr	Agric horizon	fragi	Fragipan	plac	Thin pan
al	High aluminum, low iron	fragloss	See *frag* and *gloss*	plagg	Plaggen horizon
alb	Albic horizon	fulv	light-colored melanic horizon	plinth	Plinthite
and	Ando-like	gyps	Gypsic horizon	psamm	Sand texture
anhy	Anhydrous	gloss	Tongued	quartz	High quartz
aqu	Water saturated	hal	Salty	rhod	Dark red colors
argi	Argillic horizon	hapl	Minimum horizon	sal	Salic horizon
calc, calci	Calcic horizon	hem	Intermediate decomposition	sapr	Most decomposed
camb	Cambic horizon	hist	Presence of organic materials	somb	Dark horizon
chrom	High chroma	hum	Humus	sphagn	Sphagnum moss
cry	Cold	hydr	Water	sulf	Sulfur
dur	Duripan	kand	Low-activity 1:1 silicate clay	torr	Usually dry and hot
dystr, dys	Low base saturation	lithic	Near stone	ud	Humid climates
endo	Fully water saturated	luv, lu	Illuvial	umbr	Umbric epipedon
epi	Perched water table	melan	Melanic epipedon	ust	Dry climate, usually hot in summer
eutr	High base saturation	molli	With a mollic epipedon	verm	Wormy, or mixed by animals
ferr	Iron	natr	Presence of a natric horizon	vitr	Glass
fibr	Least decomposed	pale	Old development	xer	Dry summers, moist winters
fluv	Floodplain				

Subgroup

Over 1,400 subgroups which are subdivisions of the Great group are now recognized. Classes at the subgroup level are differentiated within each Great group on the basis of properties resulting from either (i) a blending or overlapping of sets of processes in space or time that cause one kind of soil to develop from or toward another kind of soil that has been recognized at the Great group, Suborder or Order level or (ii) sets of processes or conditions that have not been recognized as criteria for any class at a higher level. A third kind of subgroup fits neither (i) nor (ii) but is considered to typify the central concept of the Great group (Ahrens & Arnold, 2000). Thus, a *Typic Hapludult* subgroup typifies the Hapludult Great group. Other subgroups may have characteristics that intergrade between those of the central concept groups. For instance, a Lithic-Reptic-Entic Hapludults subgroup are very weakly developed (shallow to bedrock and a discontinuous argillic horizon by ledges of bedrock) Ultisols, almost close to being in the Entisols Order.

Family

Soils within a given subgroup fall into a particular Family if, at a specified depth (control section), they have similar physical and chemical properties affecting the growth of plant roots. The criteria used include broad classes of particle-size, mineralogy, cation exchange activity of the clay fraction temperature and depth of the soil penetrable by roots. Table 9.5 gives examples of the classes often used.

Table 9.5:   *Some Commonly Used Particle-Size, Mineralogy, Cation Exchange Activity, and Temperature Classes Used to Differentiate Soil Families (Source: Brady & Weil, 1999)*

The characteristics generally apply to the subsoil of 50 cm depth. Other criteria used to differentiate soil families (but not shown here) include the presence of calcareous or highly aluminum toxic (allic) properties, extremely shallow depth (shallow or micro), degree of cementation, coatings on sand grains, and the presence of permanent cracks.

Particle-size class	Mineralogy class	Cation exchange activity class[b]		Soil temperature regime class		
		Term	CEC/% clay	Mean annual temperature, °C	>6°C difference between summer and winter	<6°C difference between summer and winter
Ashy	Mixed	Superactive	> 0.60	<-10	Hypergelic[c]	
Fragmental	Micaceous	Active	0.4 to 0.6	-4 to -10	Pergelic[c]	
Sandy-skeletal[a]	Siliceous	Semiactive	0.24 to 0.4	+1 ot -4	Subgelic[c]	
Sandy	Kaolinitic	Subactive	<0.24	<+8	Cryic	
Loamy	Smecttitic			<+8	Frigid[d]	Isofridgid
Clayey	Gibbsitic			+8 to +15	Mesic	Isomesic
Fine-silty	Gypsic			+15 to +22	Thermic	Isothermic
Fine-loamy	Carbonic			>+22	hyperthermic	Isohyperthermic
Etc.	Etc.					

[a]Skeletal refers to presence of up to 35% rock fragments by volume.
[b]Cation exchange activity class is not used for taxa already defined by low CEC (e.g. kandic or oxic groups).
[c]Permafrost present.
[d]Frigid is warmer in summer than Cryic.

Names of families are polynomial, often having three to five descriptive terms that modify the subgroup name. An example is fine-loamy (particle-size), mixed (mineralogy), superactive (cation exchange acativity), isohyperthermic (temperature), Typic Palehumults.

Series

Classes at the series category are the most specific units of the classification system. It is a subdivision of the family and each series is defined by a specific range of soil properties involving primarily the kind, thickness and arrangement of horizons. Features such as a hardpan within a certain distance below the surface, a distinct zone of $CaCO_3$ accumulation at a certain depth or striking slope, colour or textural characteristics greatly aid in series identification.

Series are usually given a name from some town, river or local government area either where they were first encountered or where they are dominant e.g. Agwagune silt loam, 0-2% slopes; Afikpo clay loam, 10-12% slopes, etc. There are about 19,000 soil series in the United States (Brady & Weil, 1999).

The FAO-UNESCO Soil Classification System/ World Reference Base for Soil Resources (WRB)

In 1974, the Food & Agricultural Organization (FAO) of the United Nations published its soil map of the World Legend. Compilation of the Soil Map of the World Legend was a formidable task involving the collection and correlation of soil formed from all over the world (Esu, 2005). Initially, the legend of the soil map of the World consisted of 26 first level soil grouping, comprising a total of 106 second level soil units. In 1988, a revised legend was published in which the major soil groupings in the legend were increased from 26 to 28 and the second level soil units increased from 106 to 153. Some of the main changes included the amalgamation of Lithosols, Rendzinas and Rankers into Leptosols; the split of Luvisols into Luvisols and Lixisols and, similarly, the separation of Acrisols into Acrisols and Alisols; the deletion of Xerosols and Yermosols and the introduction of Anthrosols, Plinthosols, Calcisols and Gypsisols.

In 1998, the International Union of Soil Science (IUSS) officially adopted a World Reference Base for Soil Resources (WRB) as the Union's system for soil correlation. The structure, status and definitions of the WRB are strongly influenced by the philosophy behind and experience gained with the FAO-UNESCO soil classification system. The main objective of the WRB is to provide scientific depth and background to the 1988 FAO Revised Legend, incorporating the latest knowledge relating to global soil resources and their interrelationships (ISSS-ISRIC-FAO, 1998).

Soil References Groups

Thirty "*Soil Reference Groups*" are now recognized under the WRB accommodating more than 200 ("second level") *Soil Units*. The 30 Reference Soil Groups are aggregated in 10 "sets" composed as follows:

(i)   First, a separation is made between Organic soils and Mineral soils with all Organic soils being in set #1.

(ii) The remaining (mineral) major soil Groups are each allocated to one of nine sets on the basis of "dominant identifiers", i.e. those soil forming factor(s) which most clearly conditioned soil formation.

Table 9.6 summarizes the 10 Sets, their dominant identifiers and the Reference Soil Groups within each set.

Table 9.6:   *The Ten Sets of the Reference Soil Groups of the WRB*

SETS	MAJOR IDENTIFIERS	SOIL REFERENCE GROUPS
SET #1	Organic soils	HISTOSOLS
SET #2	Mineral soils whose formation was conditioned by *human influences* (not confined to any particular region)	ANTHROSOLS
SET #3	Mineral soils whose formation was conditioned by their *parent material* - Soils developed in *volcanic material* - Soils developed in *residual and shifting sands* - Soils developed in *expanding clays*	ANDOSOLS ARENOSOLS VERTISOLS
SET #4	Mineral soils whose formation was conditioned by the *topography/physiography* of the terrain - Soils in *lowlands (wetlands) with level Topography* - Soils in *elevated regions with non-level topography*	FLUVISOLS GLEYSOLS LEPTOSOLS REGOSOLS

SETS	MAJOR IDENTIFIERS	SOIL REFERENCE GROUPS
SET #5	Mineral soils whose formation is conditioned by their *limited age* (not confined to any particular region)	CAMBISOLS
SET #6	Mineral soils whose formation was conditioned by *climate: (sub-)humid tropics*	PLINTHOSOLS FERRALSOLS NITISOLS ACRISOLS ALISOLS LIXISOLS
SET #7	Mineral soils whose formation was conditioned by *climate: arid and semi-arid regions*	SOLONCHAKS SOLONETZ GYPSISOLS DURISOLS CALCISOLS
SET #8	Mineral soils whose formation was conditioned by *climate: steppes and steppic regions*	KASTANOZEMS CHERNOZEMS PHAEOZEMS
SET #9	Mineral soils whose formation was conditioned by *climate: (sub-) humid termperate regions*	PODZOLS PLANOSOLS ALBELUVISOLS LUVISOLS UMBRISOLS
SET #10	Mineral soils whose formation was conditioned by *climate: permafrost regions*	CRYOSOLS

Table 9.7: *Diagnostic horizons, properties and materials used in the identification of Reference Soil Groups of the WRB (Source: ISSS-ISRIC-FAO, 1998)*

Horizons

Albic	light colored eluvial horizon generally associated with argic and spodic horizons
Andic	moderately weathered horizon in pyroclastic material dominated by short-range order minerals
Anthropedogenic	surface and subsurface horizons resulting from long continued cultivation
Argic	subsurface horizon with distinct clay accumulation
Calcic	horizon with accumulation of calcium carbonate
Cambic	subsurface horizon showing evidence of alteration relative to the underlying horizon(s)
Chernic	thick, well-structured, black, base-saturated surface horizon, rich in organic matter content and biological activity
Cryic	perennially frozen mineral or organic soil horizon
Duric	subsurface horizon with weakly cemented to indurated nodules cemented by silica ("durinodules")
Ferralic	strongly weathered subsurface horizon with low cation exchange capacity

Ferric	horizon in which iron is concentrated in large mottles or concretions
Folic	well-aerated, highly organic surface horizon
Fragic	natural, non-cemented subsurface horizon in which access for roots and water is restricted to interped faces.
Fulvic	thick, black horizon rich in organic matter associated with pyroclastic deposits and vegetation other than grass land
Gypsic	horizon with accumulation of gypsum
Histic	poorly aerated, waterlogged, highly organic surface horizon
Melanic	thick, black horizon rich in organic matter associated with pyroclastic deposits and grassland vegetation
Mollic	thick, dark colored surface horizon with high base saturation and moderate to high organic matter content
Natric	subsurface horizon with distinct clay accumulation and a high exchangeable sodium percentage
Nitic	clayey subsurface horizon with strongly developed, nut-shaped structure and shiny pedfaces
Ochric	weakly developed surface horizon, either light colored, or thin, having a low organic matter content
Petrocalcic	continuous cemented or indurated calcic horizon
Petroduric	continuous cemented or indurated duric horizon
Petrogypsic	continuous cemented or indurated gypsic horizon
Petroplinthic	continuous cemented or indurated plinthic horizon
Plinthic	iron-rich, humus-poor subsurface horizon which hardens irreversibly upon repeated wetting and drying
Salic	surface or shallow subsurface horizon with a high soluble salt content
Spodic	dark colored subsurface horizon with illuvial alumino-organic complexes, with or without iron
Sulfuric	extremely acid subsurface horizon with sulfuric acid resulting from oxidation of sulfides
Takyric	heavy textured crusted surface horizon occurring under arid conditions
Umbric	thick, dark colored surface horizon with low base saturationa and moderate to high organic matter content
Vertic	clayey subsurface horizon dominated by shrink-swell clays
Vitric	horizon dominated by volcanic glass and other primary minerals derived from volcanic ejecta
Yermic	surface horizon with desert pavement or a loamy vesicular crust covered by windblown deposits

Properties

Abrupt textural change	sharp increase in clay content within a limited depth range
Albeluvic tonguing	penetrations of clay and iron-depleted material into an argic horizon
Alic	very acid mineral soil material with a high amount of exchangeable aluminum
Aridic	presence of properties (low organic matter, Aeolian activity, light colors, high base saturation) in surface horizons, characteristic of arid environments
Continuous hard rock	presence of coherent rock, practically impermeable for roots
Ferralic	low cation exchange capacity
Geric	extremely low to negative effective cation exchange capacity
Gleyic	wetness producing reduced conditions caused by groundwater
Permafrost	perennial temperature at or below $0^{\circ}C$
Soft powdery lime	accumulation of translocated calcium carbonate in soft powdery form
Stagnic	wetness producing reduced conditions caused by stagnating surface water
Strongly humic	high organic matter content

Soil material

Anthropogeomorphic	unconsolidated mineral or organic material prodced by human activity

Calcaric	material containing calcium carbonate
Fluvic	fresh fluviatile, lacustrine or marine sediments
Gypsiric	material containing gypsum
Organic	material continging a very high amount of organic debris
Sulfidic	waterlogged deposit containing sulfides, and only moderate amounts of calcium carbonate
Tephric	unconsolidated, non – or only slightly weathered pyroclastic products

Table 9.8: *Key to the reference soil groups of the World Reference Base (WRB) [Source: ISSS-ISRIC-FAO, 1998]*

Soils having a *histic* or *folic* horizon
i.   *either* a. 10cm or more thick f rom the soil surface
            to a lithicor paralithic contact;
    or   b. 40 cm or mor r thick and starting with
            30 cm from the soil surface; *and*
2.   lacking an *andic* horizon starting within 30cm
    from the soil Surface
                                              HISTOSOLS

Other soils having one or more *cryic* horizons within
100 cm from the soil surface.
                                              CRYOSOLS

Others soils having *either*
1.   a *hortic, irragric, plaggic or terric* horizon 50 cm or more
    Thick; or
2.   an *anthraquic* horizon and an underlying
    hydragric horizon with a combined thickness
    e•50 cm.
                                              ANTHROSOLS

Other soils, which are *either*
1.   limited in depth by *continuous hard rock* within
    25 cm from the soil surface; or
2.   overlying material with a calcium carbonate
    equivalent e'' 40%, both within 25 cm from
    the soil surface; or
3.   containing less than 10% (by weight) fine earth
    to a depth of 75 cm or more from the soil
    surface; and
4.   having no diagnostic horizons other than a *mollic,
    ochric, Umbric, yermic or vertic* horizon
                                              LEPTOSOLS

Other soils having
1.   a *vertic* horizon within 100 cm from the soil surface;
    and
2.   after the upper 20 cm have been mixed, $\geq$ 30% clay
    in all lorizons to a depth of 100 cm or more, or to
    a contrasting layer (lithic or paralithic contact,
    petrocalcic, petroduric or *petrogypsic* horizons, between
    sedimentary discontinuity, etc.)

Other soils having
1.   *either* a *vitric* or *andic* horizon, both
    starting within 25 cm from the soil
    surface; and
2.   having no diagnostic horizons (unless
    buried deeper than 50cm) other than a
    histic, fulvic, melanic, mollic, umbric, ochric, duric
    or cambic horizon
                                              ANDOSOLS

Other soils having a *spodic* horizon starting
within 200 cm from the soil surface,
underlying an *albic, umbric or ochric* horizon,
or an *anthropedogenic* horizon less than 50 cm
thick.
                                              PODZOLS

Other soils having *either*
1.   a *petroplinthic* horizon starting within
    50 cm from the soil suface; or
2.   a *plinthic* horizon starting within 50 cm
    from the Soil surface; or
3.   a *plinthic* horizon starting within 100 cm
    from thesoil surface when underlying
    either an albic horizon or a horizon
    with *stagnic* properties.
                                              PLINTHOSOLS

Other Soils
1.   having a *ferralic* horizon at some depth
    between 25 and 200 cm from the soil
    surface; *and*
2.   lacking a *nitic* horizon *and*
3.   lacking a layer which fulfills the
    requirements of and *argic* horizon and
    which has in the upper 30 cm
    e•10% water-dispersible clay (unless the
    soil material has *geric* properties or e•1.4%
    organic carbon)
                                              FERRALSOLS

Other soils having

50 and 100 cm; *and*

3. cracks which open and close periodically.

VERTISOLS

Other soils having

1. *fluvic* soil material within 25 cm f rom the soil surface; and

2. no diagnostic horizons other than a *histic, mollic, ochric, Takyric, umbric, yermic, salic* or *sulfuric* horizon.

FLUVISOLS

Other soils having

1. a *salic* horizon starting within 50 cm from the soil surface; a*nd*

2. no diagnostic horizons other than a *histic, mollic, ochric, Takyric, yermic, calcic, cambic, duric, gypsic* or *vertic* horizon.

SOLONCHAKS

Other soils having

1. *gleyic* properties within 50 cm from the soil surface; *and*

2. no diagnostic horizons other than a *histic, mollic, ochric, Takyric, umbric, andic, calcic, cambic, gypsic, plinthic, salic* or *Sulfuric* horizon within 100 cm from the soil surface.

GLEYSOLS

Other soils having

1. a *mollic* horizon with a moist chroma of more than 2 at a depth of at least 20 cm or directly below any plough layer; and

2. concentrations of *soft powdery lime* within 100 cm from the soil surface; *and*

3. no diagnostic horizons other than an *argic, calcic, cambic, gypsic* or *vertic* horizon

KASTANOZEMS

Other soils having

1. a *mollic* horizon; *and*

2. a base saturation (by 1 M NH$_4$OAc, pH7*)* of 50% or more at least to a depth of 100 cm from the soil surface, or to a contrasting layer (lithic or paralithic contact, *petrocalcic* horizon) between 25 and 100 cm; *and*

3. no concentrations of *soft powdery lime* within 200 cm from the soil surface; *and*

4. no diagnositic horizons other than an *albic, argic, cambic* or *vertic* horizon or a *petrocalcic* horizon in the substratum.

PHAEOZEMS

Other soils having

1. either a gypsic or petrogypsic horizon within 100cm from the soil surface, *or* 15% (by volume) or more gypsum, which is accumulated under hydromorphic conditions, averaged over a depth

1. an eluvial horizon, the lower boundary of which is marked, within 100 cm from the soil surface, by an *abrupt textural change* associated with *stagnic* properties above that boundary; *and*

2. no *albeluvic tonguing.*

PLANOSOLS

Other soils having a *natric* horizon within 100 cm from the soil surface.

SOLONETZ

Other soils having

1. a *chernic* horizon or a *mollic* horizon with a moist chroma of 2 or less to a depth of a least 20 cm or directly below any plow layer; and

2. concentrations of *soft powderdery lime* starting within 50 cm of the lower limit of the Ah horizon but within 200 cm from the soil surface; *and*

3. no *petrocalcic* horizon between 25 and 100 cm from the soil surface; *and*

4. no secondary gypsum; *and*

5. no uncoated silt and sand grains on structural ped Surfaces.

CHERNOZEMS

Other soils having

1. a *nitic* horizon starting within 100 cm from the soil surface; *and*

2. gradual to diffuse horizon boundaries between the surface and the underlying horizons; *and*

3. no *ferric, plinthic* or *vertic* horizon within 100 cm from the soil surface.

NITISOLS

Others soils having

1. a base saturation (by 1 M NH$_4$OAc, pH7*)* of less than 50% in the major part between 25 and 100 cm; *and*

2. an *argic* horizon, which has a cation exchange capacity (by 1 M NH$_4$OAc, pH7*)* of less than 24 cmol$_c$ kg^{-1} clay in some part, either starting within 100 cm from the soil surface, or within 200 cm from the soil surfae if the argic horizon is overlain by loamy sand or coarser textures throughtout.

ACRISOLS

Other soils having an *argic* horizon with a cation exchange capacity (by 1 M NH$_4$OA$_c$, pH7) of less than 24 cmolc kg^{-1} clay in some throughout, either starting

of 100 cm within 1.5m from the soil surface; *and*
2. no diagnostic horizons other than an *ochric* or *cambic* horizon, an *argic* horizon permeated with gypsum or calcium carbonate, or a *calcic* or *petrocalcic* horizon underlying the gypsic horizon.

GYPSISOLS

Other soils having a *duric* or *petroduric* horizon within 100 cm from the soil surface.

DURISOLS

Other soils having
1. a *calcic* or *petrocalcic* horizon within 100 cm of the Surface; *and*
2. no diagnostic horizons other than an *ochric* or *cambic* horizon, an *argic* horizon which is calcareous, or a *gypsic* horizon underlying a petrocalcic horizon.

CALCISOLS

Other soils having an *argic* horizon within 100 cm from the soil surface with an irregular upper boundary resulting from *albeluvic tonguing* into the argic horizon.

ALBELUVISOLS

Other Soils having
1. *alic* properties in the major part between 25 and 100 cm from the soil surface; *and*
2. an *argic* horizon, which has a cation exchange capacity (by 1 M NH_4OA_c, pH 7) of 24 cmolc kg^{-1} clay or more in some part, either starting within 100 cm from the soil surface, or within 200 cm from the soil surface if the argic horizon is overlain by loamy sand or coarser textures throughout; *and*
3. no diagnostic horizons other than an *ochric, umbric, albic, andic, ferric, nitic, plinthic* or *vertic* horizon.

ALISOLS

within 100 cm from the soil surface, or within 200 cm from the soil surface if the argic horizon is overlain by loamy sand or coarser textures throughout.

LUVISOLS

Other soils having an *argic* horizon, either starting within 100cm from the soil surface, or within 200 cm from the soil surface if the argic horizon is overlain by loamy sand or coarser textures throughout

LIXISOLS

Other soils having
1. an *umbric* horizon; *and*
2. no diagnostic horizons other than an *anthropedogenic* horizon less than 50 cm thick, or an *albic* or *cambic* Horizon.

UMBRISOLS

Other soils having *either*
1. a *cambic* horizon; *or*
2. a *mollic* horizon overlying a subsoil which has a base saturation (by 1 M NH_4OA_c, pH 7) of less than 50% in some part within 100 cm from the soil surface; *or*
3. one of the following diagnostic horizons within the specified depth from the soil surface:
   a. an *andic* or *vitric* horizon between 25 and 100 cm;
   b. a *plinthic, petroplinthic* or *salic* horizon between 50 and 100 cm, in absence of loamy sand or coarser textures to a depth fo at least 100 cm.

CAMBISOLS

Other soils having
1. a texture which is loamy sand or coarser to a depth of at least 100 cm from the surface; *and*
2. less than 35% (by volume) of rock fragments or other coarse fragments within 100 cm from the soil surface; *and*
3. no diagnolstic horizons other than an *ochric, yermic* or *albic* horizon, or a *plinthic, petroplinthic* or *salic* horizon below 50 cm from the soil surface.

ARENOSOLS

Other soils.

REGOSOLS

Just as in the USDA *Soil Taxonomy*, the differentiae used in the definition or identification of Reference Soil Groups include modified versions of diagnostic horizons, diagnostic properties and features or soil materials. Table 9.7 presents a summary of definitions of the differentiate which are used. Table 9.8 also presents the key for the definitions of the thirty Reference Soil Groups of the WRB.

Simplified definitions of each of the thirty Reference Soil Groups of the WRB are listed in alphabetical order as follows:

Soil Grouping		**Description**
(1)	Acrisols	- Old, acidic clay soils with low base saturation (<50% by NH_4OAc) and a pronounced clay-enriched horizon (Bt) derived from lessivation and clay accumulation.
(2)	Albeluvisols	- Soils having an argic B horizon showing an irregular or broken upper boundary resulting from deep tonguing of the albic E into the B horizon. $\geq$
(3)	Alisols	- Soils with argic B horizons and base saturation by NH_4OAc 50% which are high in aluminium content.
(4)	Andosols	- Soils formed from materials rich in volcanic ash or glass and commonly having a dark surface horizon.
(5)	Anthrosols	- Soils in which human activities have resulted in profound modification or burial of the original soil horizons through removal or disturbance of surface horizons, cuts and fills, secular additions of organic materials, long-continued irrigation, etc.
(6)	Arenosols	- Weakly developed coarse textured (sandy) soils which possess good root permeability and water volume, but also dry out very quickly.

(7) Calcisols   -   Soils having one or more of the following: a calcic horizon, a petrocalcic horizon or concentrations of soft powdery lime within 125 cm of the surface.

(8) Cambisols   -   Weakly developed soils with a B horizon (cambic horizon) but no other diagnostic horizons.

(9) Chernozems   -   Soils with a deep, dark and humus-rich A horizon (praire soils).

(10) Cryosols   -   Soils of very cold climates that contain permafrost within 2m of the soil surface and cryoturbation processes.

(11) Durisols   -   Soils having abundant solid materials or duric or petroduric horizon within 100 cm from the soil surface.

(12) Ferralsols   -   Deeply weathered, uniformly red, yellowish red or yellow soils composed mainly of kaolinite clay, sesquioxides and gibbsite which is common in the constantly humid tropics.

(13) Fluvisols   -   Young alluvial deposits in river valleys, estuaries and coastal regions.

(14) Gleysols   -   Hydromorphic soils, often waterlogged almost throughout the year.

(15) Gypsisols   -   Soils with pronounced accumulation of gypsum ($CaSO_4$) within 125cm of the surface.

(16) Histosols   -   Organic or peat soils with organic surface layer of   40cm.

(17) Kastanozems -   Soils rich in organic matter having a brown or chestnut colour.

(18) Leptosols   -   Weakly developed shallow soils either limited by hard rock within 10cm (formerly *Lithosols*) or by highly calcareous materials or permafrost.

(19) Lixisols   -   Strongly weathered soils with argic B horizon and a base saturation (by NH_4OAc)   50%.

(20) Luvisols   -   Soils having an argic B horizon which has a CEC   24 cmol $(+)kg^{-1}$ clay and base saturation by NH_4OAc 50% throughout the B horizon.

(21) Nitisols   -   Soils with argic B horizon with nutshaped structure, shiny ped faces and appreciable amount of oxalate extractable Fe_2O_3.

(22) Phaeozems   -   Soils rich in organic matter having a dark dusky colour.

(23) Planosols   -   Soils generally developed in level or depressed relief with seasonal surface waterlogging due to an underlying slowly permeable horizon within 125cm of the surface.

(24) Plinthosols   -   Soils having 25% or more plinthite by volume in a horizon which is at least 15cm thick within 50cm of the surface or within a depth of 125 cm when underlying an E horizon or a horizon which shows gleyic or stagnic properties within 100 cm of the surface.

(25) Podzols   -   Pale, sandy soils with a bleached, white-grey E horizon and humus colloids and sesquioxides which accumulate and harden in the B horizon.

(26) Regosols   -   Soils having a mantle of loose material overlying the hard core of the earth.

(27) Solonchaks   -   Structureless saline soils with free salts often accumulated on the surface, which occur in arid regions.

(28) Solonetz   -   Highly alkaline soils (pH ≥8.5) with more than 15% exchangeable sodium saturation.

(29) Umbrisols   -   Soils with umbric epipedon. It could also have an anthropic, albic and/or a cambic horizon.

(30) Vertisols   -   Very heavy clay soils (≥30% clay) which contract and have deep, wide cracks in the dry season.

Correlation between the USDA and WRB Systems of Classification

As earlier mentioned, the USDA Soil Taxonomy system is a multiple category, hierarchical system with six categories. However, the WRB is an attempt to correlate all units of the various soils in the world and to obtain a worldwide inventory of soil resources with a common legend. As this was a cumbersome project, the legend has only two levels of soil units which are roughly equivalent to the suborder and/or Great group levels of the *Soil Taxonomy*.

An attempt has been made here in Table 9.9 to correlate the twelve Soil Orders of *Soil Taxonomy* with the 30 reference soil groups of the WRB. The author hopes that it will be understood that a closer correlation of the soil units will be achieved with the suborders and the Great groups of Soil Taxonomy.

Table 9.9:   *Approximate correlation between the Soil Orders of the USDA Soil Taxonomy and the Reference Soil Groups of the WRB*

USDA Soil Orders	Brief Description	WRB Reference Soil Group(s) Eqivalent(s)
Entisols	Recently formed soils without pedogenic horizons	Anthrosols, Arenosols, Fluvisols, Gleysols, Leptosols and Regosols
Inceptisols	Embryonic soils with few diagnostic horizons	Cambisols, Fluvisols, Umbrisols and Gleysols
Aridisols	Soils of arid regions	Calcisols, Gypsisols, Durisols Solonchaks and Solonetz
Vertisols	Dark cracking clay soils with highly expanding clay minerals	Vertisols
Mollisols	Grassland soils of $\geq$ Steppes & Prairies with black & humus-rich A horizons	Chernozems, Kastanozems, & Phaeozems
Spodosols	Soils with subsoil Accumulations of Sesquioxides & humus and bleached E horizon	Podzols
Alfisols	High base status soils with Bt horizons (Base saturation 50% by NH_4OAc	Alisols, Lixisols Luvisols, Nitisols, Plinthosols, Albeluvisols and Planosols
Ultisols	Low base status soils with Bt horizons (Base saturation < 50%).	Acrisols & Plinthosols
Oxisols	Sesquioxide rich, highly weathered and highly leached soils of the tropics	Ferralsols and Plinthosols

Histosols	Organic soils	Histosols
Andisols	Volcanic ash soils low in bulk density and high in allophone	Andosols
Gelisols	Soils of very cold climates that contain permafrost within 2m of the soil surface	Cryosols

Revision Questions

1. What is meant by Multiple Category system of soil classification? Give and demonstrate a named example.

2(a) List the twelve Soil Orders in the USDA *Soil Taxonomy* system and use only a phrase or a sentence to express the major differentiating characateristics.

  (b) Correlate each of the twelve Soil Orders with at least one reference soil group of the World Reference Base for Soil Resources (WRB).

3. Write short notes on the following concepts:
    (i)   Soil Classification
    (ii) Technical Classification
    (iii) Differentiae
    (iv) Ultisols and Oxisols
    (v)   Leptosols and Fluvisols

4. Outline the purpose and process involved in carrying out a natural soil classification.

References

1. Ahrens, R. J. & R. W. Arnold 2000. Soil Taxonomy pp. E-117-135. M. E. Summer (ed.). In Handbook of Soil Science. CRC Press, Washington, D.C.

2. Buol, S. W., F.d. Hole, R. J. McCracken & R. J. Southard. 1997. *Soil genesis and classification.* 4th Edition. The Iowa State University Press, Ames, Iowa. 527pp.

3.   Esu, I. E. 2005. Characterization, Classification and Management problems of the major soil orders in Nigeria. 26th Inaugural Lecture, University of Calabar. 65pp.

4.   FAO-UNESCO. 1988. Soil Map of the World. Revised Legend. *World Soil Resources Report 60.* FAO, Rome, 119pp.

5.   ISSS-ISRIC-FAO. 1998. World reference base for soil resources. Acco Press, Leuven, Belgium.

6.   Soil Survey Staff, 1975. Soil Taxonomy: A basic system for making and interpreting soil surveys. *USDA Handbook 436.* U.S. Govt. Printing Office, Washington DC. 754pp.

7.   Soil Survey Staff. 1994. –*Keys to Soil Taxonomy*: Sixth Edition. USDA Soil Conservation Service, Washington DC, 306pp.

8.   Soil Survey Staff. 1999. Soil Taxonomy: A basic system fo soil classification for making and interpreting soil surveys. 2nd ed. USDA – NRCS Agric. Handbook 436. U.S. Govt. Printing Office, Washington DC 869pp.

10

Principles of Soil Survey

Concept of Soil Survey

Soil survey involves the systematic examination, description, classification and mapping of soils in an area. Thus, during a soil survey, sufficient information is gathered in order to help the surveyor:

(i)   to determine the important characteristics of soils;
(ii) to classify soils into defined types and units;
(iii) to establish and to plot on maps the boundaries among kinds of soils and
(iv) to correlate and to predict the adaptability of soils to various crops, grasses, and trees as well as their behaviour and productivity under different management systems.

A soil survey also termed "soil resource inventory" in modern literature, generally consists of the soil map and the soil survey report. The *soil map* is designed to show the geographical distribution of soil types or soil mapping units in relation to other prominent physical and cultural features of the earth's surface. In this sense, *a soil mapping unit* may be defined as a kind of soil, a combination of kinds of soil, or miscellaneous land type or types that can be shown at the scale of mapping for the defined purposes and objectives of the survey. A soil survey identification legend often lists all mapping units for the survey area on the map. A soil mapping unit can contain inclusions of soils outside the limits of the taxonomic name, or names used as the name for the unit.

The *soil survey report*, on the other hand contains a full description of the natural and cultural features of the area surveyed. The characteristics, use capabilities, management requirements, predicted crop yields, and taxonomic classification for each soil type are fully described in various sections of the report. Generally, a soil map without an accompanying soil survey report cannot be useful to anyone except perhaps those soil scientists who are very familiar with the soil mapping units named in the map legend.

Types of Soil Survey

Depending on the degree of detail required, which in turn determines the intensity of observations during field mapping, and the final scale at which a soil map is to be published, soil surveys can be divided into six major types namely: Schematic Soil Surveys, Exploratory Soil Surveys, Reconnaissance Soil Surveys, Semi-Detailed Soil Surveys, Detailed Soil Surveys and Very Detailed Soil Surveys.

Schematic Soil Surveys

These are soil surveys which result in very generalized soil maps which have essentially taxonomic significance. They are often derived from intuitive extrapolation about the soils in areas where no soil surveys have taken place. Thus, such survey maps are often interpretations of climatic, geologic and vegetation maps. Most often, road-traverses linking "mapping units" are used for soil checking. Invariably, the mapping units employed in these surveys are broad, normally the Soil Orders or the Great groups of the World, and a very general picture of soil distribution pattern is given.

Generally, many national soil maps are compilations obtained through Schematic Soil Surveys and their publishing scales lie in the region of 1:1,000,000 or smaller, just like in the case of atlas maps. Indeed, a very good example of a schematic soil survey is the one that has resulted in the publication of the FAO/UNESCO Soil Map of the World at a scale of 1:5,000,000 (FAO/UNESCO, 1974).

Schematic soil surveys, also termed Syntheses or Compilations are useful in promoting public awareness about soils of a broad region. They have only some direct practical value at global and international levels of planning.

Exploratory Soil Surveys

These are soil surveys often undertaken to identify the forms of development that are physically possible within large regions of a country. The level and nature of national effort required to implement such development are assessed in general terms, providing a basis for establishing priorities and a timetable for the use of limited facilities in development.

The surveys are often undertaken at a final mapping scale of 1:250,000 to 1: 1,000,000 as general inventory maps of soils as a natural resource. At these small scales, the surveys rely on satellite imageries or extensive air-photo reconnaissance as most boundaries are drawn in the laboratory on the evidence of interpretative methods. The soil mapping units often consist of land units of various kinds, preferably enclosing identified Soil Orders and Great groups.

Reconnaissance Soil Surveys

This type of soil survey fulfills a similar purpose to those described under Exploratory surveys. However, they are often commissioned for more specific purposes rather than as a general inventory survey. Their major purpose is to identify possible areas for further intensive soil survey work as might be done before locating new irrigation or farm settlement schemes. Aerial photographic interpretation is the basis of soil boundary location, but the soil observations become slightly more numerous than in exploratory surveys and the mapping units comprise individual great groups or association of great groups.

The final maps are often published at scales of between 1: 100,000 and 1: 250,000 and are often referred to as Low Intensity Soil Surveys. Intensity of field observations (augering) are at $1km^2$ to $6.25km^2$.

Semi-Detailed Soil Surveys

Semi-detailed soil surveys are also termed Medium Intensity Soil Surveys. They are carried out to identify specific areas apparently suited to specific forms of agricultural development. A reliable interpretation is obtained of the overall proportion and general distribution of soils of differing potential for the development purpose(s) in view. Such information may be sufficient to assess the economic feasibility and even permit implementation of the less intense forms of agricultural development. However, for more intensive land utilization types such as irrigation, these surveys usually serve only a "pre-investment" purpose; to identify "project areas" within which expenditure on more intense studies for investment feasibility assessment and implementation appears to be justified (FAO, 1979).

Final maps are often published at scales of 1:25,000 to 1: 100,000 and mapping units often consist of associations of soil series, and physiographic units (enclosing identified soil series). Intensity of field observations (transects) are 300m to 500m apart.

Detailed Soil Surveys

These are High Intensity Soil Surveys which are carried out at scales between 1: 10,000 and 1: 25,000. At such scales, it becomes possible for the cartographer to indicate field boundaries upon the topographical map. Consequently, in a soil survey produced at these scales, soils can be related directly to the parcels of land which enclose them. The soil surveyor can indicate with considerable accuracy, the location of soils on the landscape and can show the intricacy of their boundaries on the map.

Detailed soil surveys are very useful in areas where soil-related problems are known to exist as one can see at a glance if a particular field is likely to have a soil with a specific problem. Also, extrapolation of information from one area to another is possible and it is also possible to indicate whether or not a development scheme is feasible. Detailed soil surveys are therefore also known as "Feasibility Surveys" (Bridges, 1982).

Soil mapping units in detailed soil surveys often consist of series or phases of soil series. Transects or grid lines in the field range from 100m to 200m apart.

Very Detailed Soil Surveys

These are Very High Intensity Soil Surveys which are carried out at scales larger than 1: 10,000. They are concerned with the precise location of high-cost projects or management problems of specialized crop production scheme. Surveys in this category would have specific objectives and the data to be collected would be contractually agreed. Boundary lines between different soil series or even phases of soil series occupy zones less than 2m wide boundaries with exactitude. Systematic field observations often on a grid are 50m to 100m apart.

Map Scales and Mapping Units

The scale of a map is the ratio of length on the map to actual length on the ground. A scale of 1: 20,000 is commonly used for detailed soil maps and indicates that 1cm on the map represent a distance of 20,000cm (or 0.2km) on the ground. The ratio is unit less; therefore, 1" on the same map represents 20,000" (or 0.316miles) on the ground.

It is pertinent to note that a *small-scale* map is one with a small scale ratio (e.g. 1: 1,000,000) on which a given object such as a 100-ha lagoon occupies only a tiny spot on the map, whereas a *large-scale* map has a large scale ratio (e.g. 1:10,000) and a similar 100-ha lagoon would occupy a relatively large part of the map.

A *map unit* is a collection of areas defined and named the same in terms of their soil components or miscellaneous area or both. Each map unit differs in some respect from all others in a survey area and is uniquely identified on a soil map. Each individual area on the map is a *delineation*. Mapping units may represent some further differentiation below the soil series level-namely, *phases* of soil series or the soil mappers may choose to group together similar or associated soils into conglomerate mapping units. Thus, the following map units could be used:

Soil phase - which is a subdivision of a soil series, which is based on some important deviation that influences the use of the soil, such as surface texture, degree of erosion, slope, stoniness or soluble salt content. Thus, an Agwagune sandy loam, 3 to 5% slopes and an Okurike sandy loam, stony phase are examples of phases of soil series.

Consociations – In a consociation, delineated areas are dominated by a single soil taxon (or miscellaneous area) of similar soils. As a rule, at least one-half of the pedons in each delineation of a soil consociation are of the same soil components that provide the name for the map unit. A consociation named for a kind of miscellaneous area is dominated by the kind of area for which it is named to the extent that inclusions do not significantly affect the use of the map unit.

Complexes and Associations – Complexes and Associations consist of two or more dissimilar components occurring in regularly repeating pattern. It is said to be a complex if the major components of the mapping unit cannot be mapped separately at a scale of 1:24,000, while the major components of an Association can be separated at a scale of about 1:24,000. In either case, the major components are sufficiently different in morphology or behaviour that the map unit cannot be called a consociation.

Undifferentiated groups – These units consist of soils that are not consistently found together, but are grouped and mapped together because their suitabilities and management are very similar for common land uses. Generally the soils are included together because some common feature such as steepness, stoniness or flooding determines use and management. Every delineation has at least one of the major components and some may have all of them.

Basic Soil Survey Procedures

Soil surveys are very rarely undertaken in this modern era of today without the assistance of aerial photography or other remote sensing data. However, in developing countries like Nigeria, where recent air photos and remote sensing data of appropriate scales are hardly available for carrying out medium to high intensity surveys, traverse survey methods involving the use of grids have been increasingly used.

Thus, it can be concluded that there are two main methods of soil survey. These include:

(a) Soil survey methods involving no aerial photo-interpretation or other remote sensing data, also called Conventional Methods and

(b) Soil survey methods involving Aerial photo-interpretation or other remote sensing data and backed by Conventional methods.

Rigid Grid Procedure

In this procedure, the survey area is divided into a number of traverses at fixed intervals along a baseline in the area to be surveyed. The intervals between traverses depends on the intensity or kind of soil survey. In high intensity surveys, the traverses are much closer to one another.

Along each of the traverses, soils are described at fixed intervals, such that the traverses and points of observation with the soil auger are fixed in a grid pattern (Fig.10.1). This method of survey gives all soils present in the area a statistically good chance of being represented on the final soil map.

The Rigid Grid procedure is particularly useful in carrying out soil surveys in landscapes covered with thick forest where visibility is severely limited. Under such conditions, position location is difficult by other means including aerial photography, and it would be difficult to use the morphological features of the landscape as a guide to the positions of soil boundaries. Instead, boundaries have to be interpolated between points of contrasting soil mapping units.

Rigid Grid surveys are best suited for high to very high intensity surveys at scales larger than 1: 10,000. Examination of the soil at fixed points throughout the survey area eliminates the subjective element of interpretation and soil boundaries are interpolated between differing observations.

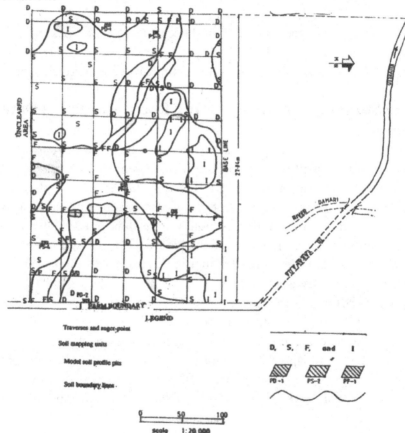

Fig. 10.1: *Base map showing baseline, traverses, auger-points and soil boundary lines in a rigid grid procedure of soil survey*

Free Survey

In this procedure, observations are also made along traverses, but the soil surveyor uses his judgment in sitting soil observations in relation to land form and other environmental features. In other words, observations are not made at fixed intervals. This method of survey is only convenient in open country where both access and visibility are almost unrestricted.

Although this is an effective and relatively rapid method of soil surveying, it can lead to a concentration of observations around the edge of soil mapping units within which "impurities" of other soils could occur. The great advantage of the method is that the surveyor is free to vary the intensity of his observations according to the intricacy of the soil pattern. This results in a greater accuracy when soil patterns are complex and does not waste time and energy when conditions are uniform.

The use of the free survey method necessitates a good base map or air photographs upon which the surveyor can work without any problems of location so that observations and boundaries are correctly placed.

Soil Survey Methods Involving Remote Sensing Data

Remote sensing involves the process of obtaining information about an object without being in physical contact with the object.

The use of air-photos in soil resources inventories began with their serving as base maps in the field, in the early 1930, when air-photos first became available in the United States of America for military purposes. Gradually, more and more techniques were developed to extract soil information from air-photos. With this method, the soil scientist can draw soil boundaries directly on the air photos. Thus, the aerial photograph serves as a base map.

Aerial photographs, which are the most common form of remote sensing data often used are contact prints made from a series of negatives of films of an aero plane flying over a piece of land usually with 55% side lap and 23% end lap. When air-photo interpretation and other remote data are used in soil mapping, it is often combined with a conventional method of soil survey. The method often involves studying intensively a portion of the survey area often termed a "sample area" by both the interpretation of the remote sensing data and cross-checking the information physically on the ground and then extrapolating the information from the sample area to the rest of the survey area.

Essentially, the steps involved are as follows:

(i)   Carry out photo-interpretation or imagery interpretation of data covering the entire area to be soil surveyed.

(ii) Select a sample area and survey it conventionally, preferably using the rigid grid procedure.

(iii) Carry out field check to correlate actual observations with photo image – observations.

(iv) Extrapolate the knowledge gained to other areas with similar photographic features or spectral signatures.

The success of the procedure involving the use of remote sensing data hinges very much on how well the sample area is selected. The following criteria for selecting a sample area in the extrapolation method should be adhered to:

(i) The sample area must be representative and for this to happen, it must have all the characteristics of the entire survey area – physiographic, climatic, vegetation, etc.

(ii) It must also include all soil units within the survey area.

(iii) It must be easily accessible.

(iv) It must be selected in such a way that the form and shape adapt to the physiographic pattern of the entire survey area.

(v) Field observations within the area must be very intensive.

(vi) A sample area should be at least 10-15% of the size of the survey area.

Phasing of Soil Survey Operations

Whatever method of soil survey employed, the following steps are usually involved in soil survey operations:

(a) Office work to plan the survey

(b) Field reconnaissance to plan field operations.

(c) Field work involving soil mapping and soil characterization.

(d) Laboratory analysis of soil samples.

(e) Cartographic work leading to the production of the soil map.

(f) Production of soil survey report and recommendations.

Office Work

Once a contract for a soil survey project has been signed, the surveyor must spend considerable time gathering information regarding the survey area.

The initial thing to do is to assemble information related to the general site information of the survey area. The surveyor must assemble road maps, geological, vegetation, topographic and land use maps of the area. If there has been a previous soil survey perhaps of a lower intensity, the soil map and soil survey report of the area should also be assembled.

Climatic data related to rainfall, evapotranspiration, temperature, relative humidity, wind, sunshine, etc, should also be assembled in the office.

If the survey is one involving aerial photo-interpretation or other remote sensing data, then the photo mosaics must be prepared and interpreted to give a general appreciation of landscapes and terrain conditions as well as a preliminary assessment of the main physiographic units requiring recognition.

All the above-mentioned preparations should culminate in the production of a base map showing the possible location of Base lines, Traverses and auger sample points.

Field Reconnaissance Visit

Having concluded preliminary office work, the next step is to pay a visit to the survey area. The purpose of the visit will be to obtain a first hand impression of the project area and be able to make arrangements for a Base Camp where the survey team will be located close to the project area. This visit will hopefully be made by the team leader or party – chief and he should make arrangements for labour, find out the wage rates, the nearest market, access roads to the survey area, the community leader of the area and generally create some awareness among the local population that such a project will soon take place in the area.

A rapid field reconnaissance should also be carried out to relate features distinguished in photographs with actual ground truth. Sufficient soil observations within recognizable physiographic units should also be carried out to identify the major diagnostic criteria to be recognized; the general level of sampling density required and to develop a preliminary working legend for soil mapping. A mapping *Legend* is a set of map unit identification names by which one can refer to areas or polygons on the map.

This initial reconnaissance should culminate in the preparation of a final budget, identification of sample areas if air photos are involved, identification of material/equipment required and the kind of transportation required for the survey.

Field Work

Field work will consist mainly of cutting and pegging of baselines and traverses, augering of soil description points, soil profile siting and description, physical measurements such as bulk density, available water capacity, porosity, etc, and infiltration measurements (if irrigation farming is envisaged). Table 10.1 lists all the materials/equipment which must be assembled to make the field trip a successful one.

Table 10.1: *Cheaklist of field material/equipment required for soil survey*

A. **Base Camp Requirements**

Beds
Mattresses
Blankets
Cooking utensils
Mosquito nets
Bed sheets
Food
Vehicles
Drums for petrol
Lanterns

B. **Baseline/Traverse Cutting**

Compasses/Theodolite
Ranging poles
Machets
Markers
Chains/Tapes
Pegs
Field Notebooks
Biros/pencils
GPS

C. **Augering Materials**

Soil Augers (type depend
on soil conditions)
Field Notebooks
Profile description sheets
Machets
Tapes (pocket)
Biros/pencils
Dil. HCI
Hydrogen peroxide
Pen Knife
Munsell colour charts

D. **Profile Pit Digging/Description**

Description sheets
Tapes
Machets
Plastic Bowls/buckets
Soil bags
Tags/Labels
Ropes
Shovels
Diggers
Scissors
Munsell soil colour charts
Geological Hammer

E. **Physical Measurements**
(**Bulk** Density/AWC/porosity)

Core sampler boxes
Hammer
Shovel
Machets

F. **Infiltration Measurements**

Double ring infiltrometers
Metal steel plates
Hammer
Recording sheets
Pocket calculators
Measuring cylinders
Rulers
Markers
Buckets/Bowls
Stop watch/clock

Laboratory Analysis of Soil Samples

Soil analysis in soil survey reports are included to define in numerical terms the physico-chemical properties of the major soils in an area. The morphological and physico-chemical properties are employed to classify the soil.General analyses often required in most soil surveys include: particle-size distribution, organic carbon, total nitrogen, available phosphorus, exchangeable bases, pH, exchange acidity, cation exchange capacity, electrical conductivity, bulk density, macroporosity,available water capacity and available micronutrient elements.

The minimum equipment needed in a soil survey laboratory to carry out the above listed analyses should include: Atomic absorption spectrophotometer, UV-visible spectrophotometer, pH meters (at least two), conductivity meters, electrical balances, hot plate with magnetic stirrers, water bath, ovens, furnaces, glasswares, flame photometer and tension plate apparatus.

Cartographic Production of the Soil Map

The principal result of a soil survey is a soil map upon which is shown the distribution of the soil mapping units (Fig. 10.2). Once the field surveyors have compiled a "field" copy of their map, it is passed on to the cartographers who have the responsibility of preparing it for publication.

One of the most critical features of a map is its scale, for this determines what can be shown, especially in terms of the smallest area capable of being represented. On a soil map, this is limited by practical considerations to an area of about $0.25cm^2$. Thus, any soil area which occupies less than $0.25cm^2$ when represented upon the map must be combined with adjacent soil mapping units.

It is pertinent to add that in virtually all soil surveys, the published map is produced at a smaller scale than that at which the field work took place. For example, the field work of a map published at 1:25,000 could be carried out at base maps of scale 1:10,000. The use of a larger scale map for field work has the advantage that the surveyor has more space in which to write his observations and when reduced to the smaller published scale, minor inconsistencies are arbitrarily removed by the cartographic process of reduction. A rule of thumb suggests that the field maps should be approximately at a scale twice the size of the published map.

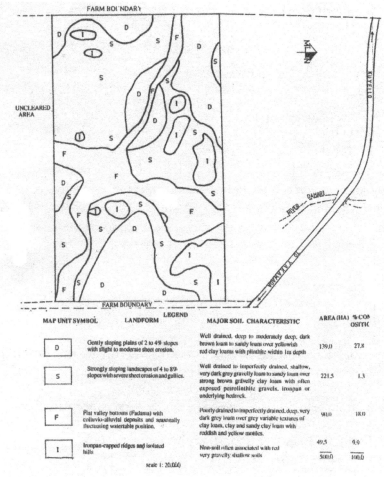

MAP UNIT SYMBOL | **LANDFORM** | **LEGEND** | **MAJOR SOIL CHARACTERISTIC** | **AREA (HA)** | **% COM OSITI**

MAP UNIT SYMBOL	LANDFORM	MAJOR SOIL CHARACTERISTIC	AREA (HA)	% COMPOSITION
D	Gently sloping plains of 2 to 4% slopes with slight to moderate sheet erosion.	Well drained, deep to moderately deep, dark brown loam to sandy loam over yellowish red clay loams with plinthite within 1m depth	139.0	27.8
S	Strongly sloping landscapes of 4 to 8% slopes with severe sheet erosion and gullies.	Well drained to imperfectly drained, shallow, very dark grey gravelly loam to sandy loam over strong brown gravelly clay loam with often exposed petrolinthic gravels, ironpan or underlying bedrock.	221.5	1.3
F	Flat valley bottoms (Fadama) with colluvio-alluvial deposits and seasonally fluctuating watertable position.	Poorly drained to imperfectly drained, deep, very dark grey loam over grey variable textures of clay loam, clay and sandy clay loam with reddish and yellow mottles.	90.0	18.0
I	Ironpan-capped ridges and isolated hills	Non-soil often associated with red very gravelly shallow soils	49.5	9.9
			500.0	100.0

scale 1: 20,000

Fig.10.2: *Soil map of Alhaji Jibrilu's farm at Kuyello, Northern Nigeria. (The map has been produced from the rigid grid Base map in Fig. 10.1)*

The Soil Survey Report

A soil survey report is a written report which accompanies the soil map when a soil survey project is undertaken. Such a report explains in detail the mapping units, their properties and relationships. According to the Soil Survey Manual (Soil Survey Staff, 1951), every soil survey report should contain:

(i) an explanation of how to use the soil map and report;

(ii) a general description of the area;

(iii) descriptions of the individual mapping units shown on the map, supplemented with Tables showing their characteristics and their relationships;

(iv)   predictions of the yields of common crops under specifically defined sets of management practices for all the soils mapped;

(v)   explanations of the management problems of each soil with special emphasis on how the characacteristics of the soil influence the problems and their solutions.

It is not always possible to meet all these requirements, but the soil survey report should go as far as possible to present such information. Usually, the form of the report varies in detail and according to the purpose it is intended to serve. In general, the analytical results are often given along with the profile descriptions of representative pedons, and often presented in the Appendix. A short account of the methods of soil analysis used and criteria for the interpretation of the results are also often included in the Appendix (Esu, 1988, 1989, 1991).

Revision Questions

1.   What is Soil Survey? Describe in an orderly sequence how you will set about conducting a detailed soil survey of a 500ha piece of land using the rigid grid method.

2.   Describe the procedure you would follow up to the point of publishing a soil map in conducting a semi-detailed soil survey of an area where no aerial photos are available.

3.   Distinguish between the following Soil Survey concepts:
    (a)   Reconnaissance and Detailed Soil Survey;
    (b)   Rigid Grid and Extrapolation Methods of soil survey.

4.   Write short notes on the following concepts:
    (a)   Semi-Detailed Soil Survey
    (b)   Soil Map and Soil Survey Report
    (c)   Rigid Grid Method of Soil Survey
    (d)   Criteria for selecting a sample area for soil survey.

References

Bridges, E.M. 1982. *Techniques of modern soil survey.* In: Bridges, E.M. and D.A. Davidson (Eds.) Principles and applications of soil geography. Longmas, London. 297pp.

Esu, I. E. 1988. *Detailed soil survey of the Tsiga scheme, Jare River Basin Irrigation Project* (Final Report) A consultancy study project commissioned by Enplan Group Consulting Engineers and Planners for the Katsina State Government of Nigeria. 82pp.

Esu, I.E. 1989. *Semi-detailed soil/land capacbility survey of the Daberam (Darugawa) irrigation project.* A consultancy study project commissioned by Parkman Nigeria Limited Consulting Engineers, Kaduna for the Katsina State Government of Nigeria. 106pp.

Esu, I. E. 1991. Detailed soil survey of NIHORT FARM at Bunkure, Kano State, Nigeria. A consultancy study project commissioned by the National Horticultural Research Institute, Ibadan, Nigeria. 72pp.

FAO, 1979. Soil Survey Investigations for Irrigation. *Soils Bulletin* No. 42, FAO, Rome. 188pp.

Soil Survey Staff, 1951. Soil Survey Manual, *Agricultural Handbook* No. 18. Soil Conservation Service, USDA. US Government Printing Office, Washington DC. 563pp.

11

Principles of Land Classification

Concept of Land Classification

Land, better referred to as "a tract of land" is defined as an area of the earth's surface, the characteristics of which embrace all reasonably stable or predictably cyclic attributes of the biosphere (vertically) above and below this area, including those of the atmosphere, soil, geology, hydrology, plant and animal populations, and the results of past and present human activity (FAO, 1976).

Land classification involves the assigning of classes, categories, or values to areas of the earth's surface (tracts of land); generally excluding water surface; for immediate or future practical use. The project, product or proposal resulting from this activity may be also generally referred to as land classification. In its broadest sense, a land classification may deal with land use, land evaluation, land systems, land capability, land inventory, and terrain evaluation. It is also concerned with soil survey, soil survey interpretation, and soil capability, suitability or limitations. Many kinds of resources inventories such as vegetative, climatic, geologic, topographic, hydrologic, economic, sociologic and demographic surveys, relate to land classification (Olson, 1974).

When land classification is properly carried out, it can organize knowledge of extremely complex landscape features into units that can be readily understood and thereby manipulated by man. Land classification generally involves two parts or phases: resource inventory and analysis and categorization. Generally, the first consists of gathering data and delineating land characteristics on maps (e.g. soils, slopes and vegetation); these data describe the land resource as it exists at the time of the survey.

The analysis and categorization, on the other hand, put the basic data into a form that can be used, generally for a specific purpose. The land classification is the part of the inventory and evaluation that makes the data relevant and applicable to the problem at hand; usually it is created explicitly to solve a specific problem.

Thus, the standard soil survey map just discussed in chapter 10, shows the different kinds of soil that are significant and their location in relation to other features of the landscape. These maps are intended to meet the needs of users, with widely different problems such as those involved in agriculture, engineering construction, etc., therefore, they contain considerable detail to show important basic soil differences. The information on the soil map must therefore be explained in a way that has meaning to the user. Soil maps can be interpreted by:
(1)   the individual kinds of soil on the map and
(2)   the grouping of soils that behave alike in response to management and treatment (Esu, 1988) (Fig. 11.1).

Soils are grouped in different ways according to the specific needs of the map user. The different ways of grouping soils has been accomplished through several systems of land capability/suitability classification methods; some of these are discussed below:

Land Capability Classification System

This system was developed by the Soil Conservation Service of the United States Department of Agriculture (Klingebiel and Montgomery, 1966). The land capability classification is based on the detailed soil survey, generally published at scales of 1:10,000 to 1:20,000 in the United States.

The classification consists essentially of grouping the various soil mapping units "primarily on the basis of their capability to produce common cultivated crops and pasture plants over a long period of time" (Klingebiel and Montgomery, 1966).

The capability groups are made at three levels of management: Land capability class; land capability subclass and land capability unit.

Land Capability Classes

There are eight land capability classes (I-VIII) which range from the best and most easily farmed land (Class I) to land which has no value for cultivation, grazing or forestry, but may be suited for wildlife, recreation or for watershed protection (Class VIII). They all fall into two broad groups of land; one suited for cultivation (Class I-IV); the other not suited for cultivation (Class V-VIII).

Classes I, II and III include land suited for regular cultivation. Class I land is very good, nearly level land, with deep, easily worked soils that can be cultivated safely with ordinary good farming methods. Class II land is good but it has some limitations. It needs moderately intensive treatment if it is to be cultivated safely. Such treatment for example, may be contouring and cover cropping to control erosion or simple water-management operations to conserve rainfall. Class III land is moderately good and can be used for cultivated crops regularly in a good rotation, if ploughed on the contour on sloping fields. It has limitations of such degree that intensive treatment is necessary. The treatment may be terracing or strip cropping to control erosion, or intensive water management on flat, wet areas.

Class IV is fairly good land, but its use for cropping is very limited by natural features such as slope, erosion, adverse soil characteristics, or adverse climate. As a rule, its best use is for pasture, but some of it may be cultivated occasionally with proper safeguards.

Class V, VI, and VII are not suited for any cultivation but may be used for grazing or forestry, according to adaptability. Class V land has few natural limitations for such use, and needs only good management. Class VI needs protective measures, usually because of slope or shallow soil. Class VII needs extreme care to overcome or cope with its major limitations, which usually are steep slope, very shallow soil, or other very unfavourable features.

Class VIII land is suited for wildlife, recreation or watershed protection. It is usually characterized by such features as extreme steepness, roughness, stoniness, wetness, sandiness or erodibility. These characteristics make it unfit for any safe or economic cultivation, grazing, or forestry.

Land Capability Subclasses

The land subclasses represent convenient groupings of subordinate characteristics within a land capability class. Reasons for placing certain soil mapping units in a class lower than Class I are often indicated by adding the letters e, w, s and c as suffixes, singly or in combination to the class number to show the nature of the deficiencies or limitations which has necessitated the down grading of a land class from Class I to say

Class III. For example, a land subclass IIIe, has limitation due to severe wind or water erosion, while a land subclass IIIs is one with land limited chiefly by soil conditions.

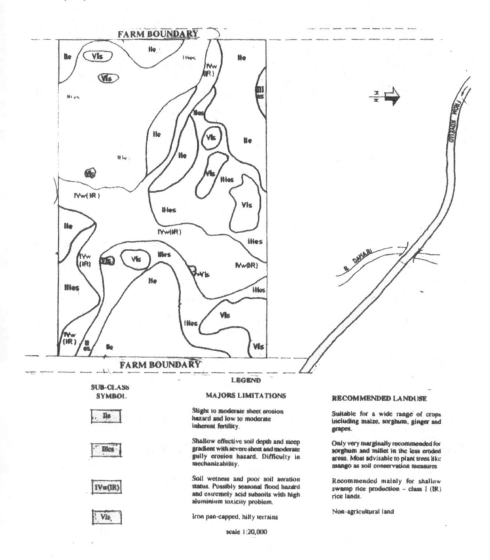

Fig. 11.1: *Land capability classification map derived from the soil map in Fig. 10.2*

Thus, the eight land classes express degree of usefulness. The subclasses express kinds of limitations within each class. The four subclasses, which may be recognized in most of the land capability classes except Class I, are:

e   -   land dominantly subject to wind or water erosion, or both

w   -   land subject to presence of excess wetness or to overflow

s   -   land limited chiefly by soil conditions, such as excessive sandy texture, excess gravel or stones, or shallow depth

c   -   land limited chiefly by climate, either inadequate precipitation or low temperature.

Land Capability Units

Within each subclass, the lands that are suited for essentially the same kind of management and the same kind of conservation treatment is designated as a land capability unit. A land capability unit is essentially uniform in all major characteristics that affects its management and conservation. It is the smallest unit recognized in the land-capability classification system and may be likened to a soil phase – within a soil series.

Land capability units are designated by ordinary Arabic numerals as IIe-1, IIe-2, IIIe-1, IIIes-1, IIIes-2, etc.

Land Suitability Classification Systems

Concept of Land Suitability Classification

Land suitability classification is an interpretative grouping made primarily for agricultural purposes. Land suitability has been defined as the fitness of a given tract of land for a specified kind of land use (FAO, 1976). Thus, for instance, it is possible to map land suitability for rainfed maize, sorghum or for surface irrigation in general or for sprinkler irrigated sugar cane. This is in contrast to the USDA land capability classification system already discussed, which is a much more broad classification of soils for widely defined land uses and management systems such as agriculture, forestry, wildlife and aesthetic appreciation.

According to Verheye (1986), land suitability classification evaluation involves six basic steps which include:

(i)   Defining the land utilization type(s) i.e. a specific subdivision of major kind of land use e.g. rainfed maize production.

(ii)   Establishing the crop requirements (agro- climatic, soil and physiography) or land qualities i.e. definitions of suitable land qualities and characteristics.

(iii) Collection of data on these environmental factors enumerated in (ii) above i.e. natural resource surveys.

(iv)   Matching (comparison) of data with requirements resulting in provisional suitability classes.

(v)   Checking with existing land use pattern and with yield data in relation to suitability classes.

(vi)   Final definition of suitability classes (actual and potential).

Land Suitability Classification Involving Rainfed Agriculture

The land suitability classification system commonly recommended for rainfed agriculture is the system proposed by the FAO (1976, 1984) commonly called the "Framework for Land Evaluation". The system involves a suitability classification with four categories: Land suitability Order, Class, Subclass and Unit. The structure of the classification is summarized in Table 11.1, while the definitions of suitability Orders and Classes are presented in Table 11.2.

There are two Orders, "Suitable" (S) and "Not Suitable" (N) for the use under consideration. A land is considered unsuitable if proposed use is technically impracticable e.g. cultivating very shallow (rocky) soils is environmentally undesirable as use may lead to severe erosion or is economically unprofitable. Within the Order "Suitable" (S) there are three classes, indicating the degree of suitability – highly (SI), moderately (S2) and marginally (S3) suitable. In the 'Not Suitable' (N) Order, there are two classes; currently not suitable (NI) i.e. can be made suitable by new technology or more costly modifications and permanently not suitable (N2)

Table 11.1: *Structure of the FAO Land Suitability Classification (Adapted from FAO Soils Bull. 52, 1984)*

Order	Category		
	Class (1)	Subclass (2)	Unit (3)
Kind of suitability	degree(s) of (un) suitability	Kind of limitation(s)	Management requirements of production characteristics

S Suitable
- S1 Highly suitable
- S2 Moderately Suitable
  - S2 n
  - S2 e
    - S2 e-1
    - S2 e-2
    - S2 e-3
  - S2m
  - etc
    - etc
- S3 Marginally Suitable
- etc

Sc Condiitonally Suitable (4)
- Sc 2
- etc
  - Sc 2 m
  - etc

N Not suitable
- N1 Currently Not suitable
  - N1 m
  - N1 e
  - etc
- N2 Permanently not suitable
- etc

Notes:

(1) The class names shown here are those recommended for a three-class system. Up to five suitable classes are permitted.

(2) The number of subclasses should be kept to the minimum necessary to distinguish land with significantly different management requirements by production potential. As few limitation symbols as possible should be used for each subclass. Note that S1 land is not divided into subclasses.

(3) Units are normally for use at the farm planning level, and are often defineable by differences in detail of their limitation(s).

(4)   For small areas of land where certain conditions to those specified for the 'suitable' classes must be fulfilled for successful land use; once these conditions are met, the land is included in the class or subclass indicated by the code following the Sc designation. NB: 'conditionally suitable' does not imply that the interpretation is uncertain, either because the land is only marginally suitable or because the relevant factors are not understood.

Table 11.2: *Definitions of Land Suitability Orders and Classes (Adapted from FAO Soils Bull, 52, 1984)*

Categories	Definitions
Suitability Orders	
Order S. Suitable	Land on which sustained use of the kind under consideration is expected to yield benefits which justify the inputs, without unacceptable risk of damage to land resources.
Order N, Not suitable	Land which has qualities that preclude sustained use of the kind under consideration.
Suitability Classes	
Class SI, highly Suitable	Land having no significant limitations to sustained application of a given use, or only minor limitations that will not significantly reduce productivity or benefits and will not raise inputs above an acceptable level.
Class S2, moderately Suitable	Land having limitations which in aggregate are moderately severe for sustained application of a given use: the limitations will reduce productivity or benefits and increase required inputs to the extent that the overall advantage to be gained from the use, although still attractive, will be appreciably inferior to that expected on class S1 land.
Class S3, marginally Suitable	Land having limitations which in aggregate are severe for sustained application of a given use and will so reduce productivity or benefits, or

Categories	Definitions
	increase required inputs, that this expenditure will be only marginally justified.
Class N1, currently not suitable	Land having limitations which may be surmountable in time but which cannot be corrected with existing knowledge at currently acceptable cost; the limitations are so severe as to preclude successful sustained use of the land in the given manner.
Class N2, permanently not suitable	Land having limitations which appear so severe as to preclude any possibilities of successful sustained use of the land in the given manner.
NR, not relevant	Land which has not been assessed for a given use, because the application of the use to that area is precluded by the initial assumptions of the evaluation.

The highly suitable calss (S1) does not indicate perfection but relative quality while the difference between S3 and N1 varies with relative costs and prices over time. The symbol NR (not relevant) is used where the basic assumptions of the evaluation excludes this portion of land from the use under consideration.

Suitability subclasses indicate kinds or limitations e.g. moisture deficiency, erosion hazard. They are indicated by lower-case letters placed after the class symbol, e.g. S2m, S2c. There are no subclasses for S1. There is no limit to the number of limitations that can be recognized in a survey, and some of the limiting land qualities and the symbols often used as suffixes to indicate them are presented in Table 11.3.

Table 11.3: *Suggested Letter Suffixes for Indicating Land Suitability Subclasses (Adapted from FAO Soils Bull, 52, 1984).*

	Limiting Land Quality	Letter Suffix
1	Radiation regime (sunshine)	u
2	Temperature regime	+ c
3	Moisture availability	+ m
4	Oxygen availability to roots (drainage)	+ w
5	Nutrient availability	+ n
6	Nutrient retention	n
7	Rooting conditions	+ r
8	Conditions affecting germination and establishment	g
9	Air humidity as affecting growth	h
10	Conditions for ripening	i
11	Flood hazard	+ f
12	Climatic hazards	c
13	Excess of salts	+ z
14	Toxicities	+ x
15	Pests and diseases	p
16	Soil workability	k
17	Potenital for mechanization	+ q
18	Land preparation and clearance requirements (vegetation)	v
19	Conditions for storage and processing	j
20	Conditions affecting timing of production	y
21	Access within the production unit	a
22	Size of potential management units (bigness)	b
23	Location	l
24	Erosion hazard	+ e
25	Soil degradation hazard	d

Note: The most commonly used land quality symbols are marked +
The letters n and c are used for two related land qualities.

Suitability units are subdivisions of subclasses differing in detailed aspects of their production characteristics or management requirements. They are numbered successively e.g. S2e-1, S2e-2. For the common case of land suitability evaluation for specific crops under rainfed agriculture, supplementary definitions of suitability classes have been devised in terms of expected crop yield and inputs required. However, there are at present very few land qualities for which the quantitative information needed to apply these criteria precisely is available (Dent & Young, 1981).

Land Suitability Classification Involving Irrigated Agriculture

The system designed by the United States Department of Interior, Bureau of Reclamation, the USBR system (1953) and modifications of it are used widely for irrigation suitability evaluation. Unlike many other systems, they are explicity based on the economics of land development although physical features are used as a basis for the economic rating. Depending on the nature and scale of the surveys, such systems can have varying degrees of quantitative assessment built into both the physical and the economic criteria used.

The basic system is intended to reflect the 'payment capacity' of the land to support agricultural production. Ideally, farm budgets are used for evaluating the costs and benefits, initially for the best land (Class I). Less suitable land is then downgraded according to the economic effects of a number of physical land deficiencies acting singly or together; these are considered under the basic headings 'soil', 'topography' and 'drainage'. There are four main classes, as follows:

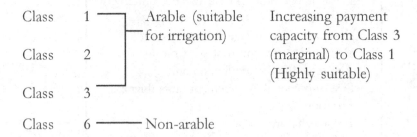

Class	1	Arable (suitable for irrigation)	Increasing payment capacity from Class 3 (marginal) to Class 1 (Highly suitable)
Class	2		
Class	3		
Class	6	Non-arable	

For detailed studies, two additional special classes are recognized.

Class   4:   Suitable for special use of restricted crop range.

Class   5:   Temporarily non-arable (pending further study of problems).

It should be noted that the payment capacities of these last two classes are *not* necessarily intermediate between those of Classes 3 and 6, but will depend on the particular economics of a given project. The crops included in the restricted range of Class 4 may have a higher or lower payment capacity than those of the other 'arable' classes, and the problems of Class 5 lands (e.g. behaviour of regional ground water table) may prove to be anything between negligible and completely insurmountable at reasonable cost.

The general class descriptions of the USBR system (Table 11.4) are invariably adapted for particular projects. For example in South-East Asia (Laos & Thailand) two separate classes for rice land, which would otherwise be Class 4 are recognized; Classes 1R and 2R, and another two arable land classes for diversified crop production are also used (Table 11.5).

In projects with little or no economic data, the classes may be defined on the basis of physical criteria, with implied economic consequences. One such modified USBR system used by a local consultancy group is presented in Table 11.6.

Each classification on the USBR system is thus usually a relative classification depending on the specific use(s) or crops envisaged, the economic circumstances of the proposed development project at the time of survey, and the inputs required to bring the land to full productive capacity. A distinction needs to be drawn, therefore, between land preparation costs (e.g. of land reclamation, clearance and grading; installation of irrigation or drainage systems) and operational costs (e.g. fertilizer applications, tillage) and also between 'correctable' and 'non-correctable' land deficiencies (e.g. minor undulations, salinity and pH problems, as against steep slopes, soil depth and texture problems). However, the scope of practicable changes will vary depending on the economic feasibility of ameliorative processes within each project.

Table 11.4: *Example of Modified USBR Land Suitability Classes*

Class I: Irrigable

Lands that are most suitable for irrigation farming, being capable of producing sustained and relatively high yields of a wide range of climatically suited crops at reasonable cost. They have gentle slopes. The soils are deep and of medium to fairly fine texture with mellow open structure allowing easy penetration of roots, air and water, and having free drainage – yet good available moisture capacity. These soils are free from harmful accumulations of soluble salts. Both soils and topographic conditions are such that no specific farm drainage requirements are anticipated, minimum erosion will result from irrigations and land development can be accomplished at relatively low cost.

Class 2: Irrigable

This class comprises lands of moderate suitability for irrigated farming, being measurably lower than Class 1 in productive capacity, adapted to a somewhat narrower range of crops, more expensive to prepare for irrigation or more costly to farm. They may have a lower available moisture capacity, as indicated by coarse textures. Any one of the limitations may be sufficient to reduce the lands from Class 1 to Class 2, but frequently a combination of one or more of them is operating.

Class 3: Irrigable

Land that are suitable for development, but are approaching marginality for irrigation and are of distinctly restricted suitability because of more extreme deficiencies in the soil, topographic or drainage characteristics than described for class 2 lands. They may have restricted crop use, requires larger amounts of irrigation water or special irrigation practices, and demand greater fertilization or more intensive soil improvement practices. They may have uneven topography or restricted drainage amendable to correction but only at relatively high costs. Generally, greater risk may be involved in farming Class 3 lands.

Class 4: Restricted irrigable

Lands are included in this class when a specific excessive deficiency or deficiencies indicate that the land is irrigable only for particular types of crops, or by restricted methods of irrigation. The deficiency may be very slow profile permeability and poor drainage, leading to a restriction to rice, steeply sloping land whose utility seems best served by tree crops, or very coarse soils with high infiltration rates which are recommended for irrigation by overhead methods. The Class 4 lands have a range of payment capacity which, in some circumstances, may be as high as or higher than in the irrigable land classes.

Class 5: Provisionally non-irrigable

Land in this class are not considered economically irrigable pending further study. They consist, for example, of land underlain by laterite at depths between 1.0 and 1.5m. Final designation of land class depends on detailed assessment of the hazard presented by the laterite, particularly as related to regional drainage.

Class 6: Non-Irrigable

Lands in this class include those failing to meet the minimum requirements of the classes described above. They comprise land with excessive slopes or dissection; with inadequate drainage and very variable soils, or with a liability to flooding. Under the envisaged development programme, these lands are not considered to have sufficient repayment capacity to warrant irrigation.

(Adapted from USBR, 1953)

Table 11.5: *Modified USBR Land classification specification used in Laos and Thailand (Source: FAO, 1979)*

Land characteristics	For diversified crop production		For wetland rice production	
	Class 1 – Arable	Class 2 – Arable	Class IR Arable	Class 2R Arable
Soils				
Texture				
Surface, 0-30cm	Fine sandy loam to clay loam	Loamy fine sand to permeable clay	Fine sandy loam to clay	Loamy sand to clay
Subsurface	Sandy loam to permeable clay	Loamy fine sand to permeable clay	Loamy sand to clay	Sand to clay
Depth (after land development) to: clean sand or gravel pisolites in	> 90 cm	> 60 cm	> 60 cm	> 30 cm
permeable matrix	> 90 cm	> 60 cm	> 60 cm	> 30 cm
permeable armour	> 150 cm	> 90 cm	> 60 cm	> 45 cm
relatively impermeable zone (water)	> 210 cm	> 210 cm	> 210 cm	> 210 cm
Available water capacity	15 cm or more in 120 cm depth within 2.5 cm in 0.30 cm	8 cm or more in 120 cm depth within 2.5 cm in 30 cm	not applicable	not applicable

Table 11.5: *(Contd)* For diversified crop production For wetland rice production

Land characteristics	Class 1 – Arable	Class 2 – Arable	Class IR Arable	Class 2R Arable
Reaction				
pH in 0.01 M CaC_2	>5.0 < 7.7	> 4.0 < 8.0	> 5.5	> 5.0 may be less provided aluminium active ion levels are satisfactory
pH in H_2O (1:1)	> 5.5 < 8.2	> 4.5 < 8.5		
pH in (anaerobic)				
Acidity [1]				
Neutral salt exchangeacidity	None	May be moderate		
Buffered salt exchange acidity	May be moderate	May be moderate		
Anion exchange acidity	May be moderate	May be moderate		
Inorganic (acid Sulphate soil)	None	None	None	None
Exchangeable sodium Percentage	< 20	< 20	< 20	< 20
Sodium-adsorption-ratio (soil solution)				

[1] Appraisal is dependent on charge characteristics and ion populations as related to cropping pattern.

Table 11.5: (Contd)

Land characteristics	For diversified crop production		For wetland rice production	
	Class 1 – Arable	Class 2 – Arable	Class IR Arable	Class 2R Arabe
Cation-exchange capacity(at soil pH or surface Soil, (0-30cm)	> 10 meq/100g	> 5 meq/100g	> 10 meq/100g	> 3 meq/100g
Base status: Calcium Magnesium Potassium Sodium	> 2.0 meq/100g > 1.0 meq/100q > 0.4 meq/100g < 0.2 meq/100g			
Reduction/products active iron Soil solution (after Prolonged flooding)			< 200 ppm	< 500 ppm
Salinity (at equilibrium under irrigation) Electrical conductivity Soil solution	saturation extract < 4.0 mmhos/cm	< 10.0 mmhos/cm	< 4.0 mmhos/cm	<8.0 mmhos/cm<
Topography Slope	< 2% > 0.25%	< 5% > 0.25%	< 2%	5%

Table 11.5: *(Contd)*

Land characteristics	For diversified crop production		For wetland rice production	
	Class 1 – Arable	Class 2 – Arable	Class IR Arable	Class 2R Arable
Drainage				
Flooding	None	None		
Internal	Good	Good	Very slow	Slow

Class 5s – Unresolved Potential, Diversified Cropland

Includes lands having unresolved potential for irrigation development involving only diversified cropping. These lands meet all the requirements of arability for diversified cropland, Class 2, except for effective soil cation exchange capacity and have soil textures of fine sandy loam or finer.

Class 5 – Tentatively Non-Arable

Includes lands which will require additional economic and engineering studies to determine their irrigability. This designation (5) is particularly suited to areas above proposed canal lines pending the determination of feasibility of service. Also applies to suspected high or isolated lands within the known service area and lands subject to seasonal inundations requiring project flood protection works.

Class 6 – Non-Arable

Includes lands which do not meet the minimum requirements for the other land classes and are not suitable for irrigation. They include very shallow soils, salt affected soils that are reclaimable with difficulty, etc.

Table 11.6: *Example of modified USBR land suitability class specifications for gravity irrigation used by a local consultancy group with only physical parameters involved (Source: Esu, 1988)*

Land Characteristics	Class 1 – irrigable	Class 2 – Irrigable	Class 3 – Irrigable
Soil			
Topsoil texture (0 to 30 cm)	Porous fine sandy loams to fine sandy clay loams	Fine sand to loamy fine sand	Fine sand to loamy fine sand
Subsoil texture (30 to 80 cm)	As topsoils	Porous fine sandy loams to fine sandy Clay loams	Fine sand to loamy fine sand
Effective Depth	> 150 cm	> 150 cm	> 150 cm
Available water Capacity	> 150mm m^{-1} soil	150-120mm m^{-1} soil	120-90mm m^{-1} soil
Infiltration (IR) after 4 hours	0.7-5.0cm h^{-1}	5.0-12.0cm h^{-1}	12.0-15.0cm h^{-1}
Topography			
Slopes	< 0.5^0	< 0.5^0	0.5° to 1°
Levelling requirements	<350 m^3ha^{-1}	350-750m^3 ha^{-1}	750-1 000 m^3 ha^{-1}
Vegetation cover	Moderate to low land clearing costs	Moderate land clearing costs	Moderate to high land clearing costs

Table 11.6 (*Contd.*)

Land Characteristics	Class 1 – irrigable	Class 2 – Irrigable	Class 3 – Irrigable
Drainage			
Ground water table	Normally ≥10 m	7-10 m	5-7 m
Drainage	No immediate farm drainage required; Profiles well drained.	No immediate farm drainage required; profiles well drained.	Minor farm drainage required in places. Good to moderate profile rainage.

Class 4: Restricted irrigable or special use
Includes lands with coarse soils (fine and medium sands, loamy fine sands) to depth; high IR rates of > 15.0 cm h^{-1}; low AWC values; slopes between 1 and 3: land leveling requirements > 1,000 m^3/ha^{-1}: GWT levels within 5 m of the surface; poorly drained profiles. These soils are considered suitable only for overhead or drip irrigation systems, although small basin irrigation may be possible on a small scale.

Class 5: Provisionally non-irrigable
Includes lands underlain by laterite within 150cm of the soil surface; additional economic and engineering studies are required to determine whether drainage is required or is practical.

Class 6: Non-irrigable
Includes lands with excessive topographic, flooding or drainage problems which are considered to be non-correctable at an economic rate.

USBR terminology also distinguishes between 'arable' land and 'irrigable' land. The former is land that is mapped as suitable for irrigation, but the latter is land actually selected for development, so land out of command and small, isolated or odd-shaped tracts are excluded.

Non-Parametric Methods of Land Suitability Classification for Rainfed and Irrigated Agriculture

Apart from the land suitability classification systems for rainfed (FAO, 1976; 1984) and irrigated agriculture (USBR, 1953) just discussed, there are numerous soil or land productivity rating models developed by Soil Scientists working in various parts of the world, and which are often adapted to the local environments in which they are working. A few of such indicies of productivity ratings will be discussed here.

Riquier (1971) developed a model of rating soil productivity based on the index of productivity (IP). According to his procedure, index of productivity is determined from nine soil factors, each factor being rated on a suitability scale from 0 to 100. The resultant index is obtained by multiplying the actual ratings of the different factors:

$$IP = H \times D \times P \times T \times or \; \frac{N}{S} \times O \times A \times M$$

Where H = Soil moisture; D = drainage conditions; P = effective soil depth; T = texture/structure; N = base saturation; S = soluble salts; O = Organic matter contents; A = nature/CEC of clay mineral; M = mineral reserve.

According to the resulting index of productivity, the soil is assigned to one of five productivity classes:

Class 1 =	excellent with rating	65 – 100
Class 2 =	good, with rating	35 – 64
Class 3 =	average, with rating	20 – 34
Class 4 =	poor, with rating	8 – 19
Class 5 =	extremely poor to nil with rating	0 – 7

A unique suitability soil rating system, the Storie Index, was developed for use in southern California (Storie, 1937), but is adaptable to many other arid and semi-arid regions. The four soil characteristics used in the soil-plant rating scale are:

Factor A - Profile characteristics that influence the depth and quality of the root zone.

Factor B - Textural class of the surface soil as it relates to infiltration, permeability, water retention capacity and ease of tillage.

Factor C - Slope as a soil-plant limitation related to irrigation potential.

Factor X - Other characteristics other than those in Factor A, B and C which limit plant growth, including poor drainage, excessive erosion, excessive salts, high exchangeable sodium, low inherent fertility and/or high acidity, which are multiplied separately for input into the main equation.

Factors A, B, C and X are individually assigned a rating with a maximum of 100 percent. The composite soil index rating is determined by multiplying the four separate rating as decimal fractions of the percentages of each of the factors (Factor A x Factor B x Factor C x Factor X). According to the resulting index, the soil is assigned to one of six productivity ratings for intensive agriculture as tabulated below:

Composite soil	Soil grade for intensive agriculture	
Rating (Storie-Index)	Numerical	Relative
80 – 100%	1	Excellent
60 – 80%	2	Very good
40 – 60%	3	Good
20 – 40 %	4	Moderately good
10 – 20%	5	Poor
Less than 10%	6	Very poor

Esu (1982) advocated the use of a multiplicative soil rating index based on a modified version of the Storie-Index for selecting irrigable lands in Nigeria and designed a chart for rating the soils within the Kaduna area of Nigeria. Following his procedure, he rated the dominant soil units in a typical toposequence within the Kaduna area for irrigation suitability of diversified crops under gravity irrigation as summarized in Table 11.7.

Productivity index rating methods, with suitable refinements to suit local environments, give useful and reproducible results in specific areas and circumstances, however, they are not usually recommended for universal application, since widely inaccurate predictions can arise when a productivity index is used outside the area where it was developed. A major drawback of the system is the use of several arbitrary procedures arising from subjectivity in the selection of properties to be used, the formulation of the equation, the numerical value to be assigned to each range of property and the translation into planning or operational terms of the productivity index obtained.

Table 11.7: Soil Index rating for major soil units in the Kaduna area of Nigeria (Source: Esu, 1982)

Taxonomic Classification		Topographic Position	Drainage Class	Soil grade for irrigated agriculture		
USDA Soil Taxonomy	FAO/UNESCO			Numerical	Percentage	Relative
Oxic Paleustalf, fine – loamy, mixed, isohyperthermic	Dystric Nitosols	Summit to shoulder	Well-drained	3	51	Good
Plinthustalf, fine-loamy mixed, isohyperthermic	Plinthic Luvisols	Back slope to Toe slope	Well-drained	4	34	Moderately good
Typic Tropaquept, fine, kaolinitic, isohyperthermic	Dystric Gleysols	Foot slope	Poorly drained	5	10	Poor
Typic Ustifluvent, coarse-loamy, mixed, isohyperthermic	Eutric Fluvisols	Levees	Moderately well-drained	1	81	Excellent

Note: Ratings were for the irrigation of diversified arable crops (not wetland crops) under gravity irrigation.

Revision Questions

1. Distinguish clearly between the concepts of "soil" and "land". What do you understand by land classification?

2(a) Outline the role of soil survey in land suitability classification evaluation. Use diagrams to illustrate your answer.
  (b) What are the main features of the USDA land capability classification system?
  (c) Use a named example to illustrate your understanding of the term "Soil Productivity Rating".

3(a) Outline the basic steps involved in a land suitability classification evaluation.
  (b) Discuss the structure of the FAO land suitability classification system and state how it differs from the USBR *irrigation land suitability* classification system.

4. Physico-chemical properties of floodplain soils in the Hadejia valley are presented in Table 11.8. Use the data as a guide in answering the following questions.
  (a) Comment exhaustively on the fertility status of each of the three soils.
  (b) Do you consider the soils saline or sodic? If so, what steps will you take to ameliorate any of such conditions?
  (c) Which of the soils are likely to be calcareous? What are the likely adverse effects this might have on such a soil when irrigated?
  (d) Carry out a land suitability classification ratings of the soils for sorghum and swamp rice production under rainfed conditions using the FAO system and under irrigated conditions using the USBR system.

Table 11.8: *Physico-chemical properties of Hadeija floodplain soils*

Horizon	Depth	Particle size distribution			pH(H2O)	Organic C	Total N	Avail. P	EC	Exchangeable bases				CEC
		Sand 2000-50µm	Silt 50-2µm	Clay <2µm						Ca	Mg	K	Na	
	cm	%				%	%	mg/kg	dSm-1	meq/100g soil				
Levee soils														
Ap1	0-15	54	24	22	6.1	1.12	0.12	10.15	1.20	5.5	2.2	0.18	4.60	12.8
Ap2	15-45	54	22	24	8.5	0.24	0.06	4.20	0.84	5.5	2.0	0.24	5.70	12.0
Cg	45-110	56	22	22	8.4	0.16	0.04	2.45	0.55	3.7	1.3	1.10	5.60	9.1
2Cg1	100-140	74	20	6	8.3	0.08	0.02	3.30	0.19	2.5	0.5	0.05	1.30	6.4
2Cg2	140-175	80	8	12	7.7	0.12	0.02	1.40	0.20	3.2	1.0	0.08	1.50	8.1
Backswamp soils														
Apg	0-20	20	22	58	6.1	4.21	0.48	2.80	0.25	18.2	7.9	1.33	1.40	33.8
ACg	20-70	10	16	74	7.6	0.82	0.10	4.20	0.16	17.5	7.8	1.23	3.50	31.2
Cg	70-130	12	30	58	7.9	0.22	0.05	2.80	0.20	19.9	6.5	0.92	3.30	29.4
Terrace soils														
Apg	0-6	34	47	19	4.9	0.66	0.05	3.85	0.10	3.1	1.4	0.22	1.09	8.4
Eg	6-35	34	47	19	5.3	0.36	0.02	1.40	0.20	5.4	1.5	0.05	2.39	9.8
2Btg1	35-80	32	37	31	6.3	0.04	0.01	2.80	0.28	9.4	2.2	0.11	3.70	15.9
2Btg2	80-110	16	51	33	7.1	0.14	0.01	2.63	0.28	9.9	2.3	0.11	4.02	16.8
2Cg	110-160	32	49	19	7.9	-	-	2.63	0.20	4.9	1.1	0.06	2.94	10.4

5.   A detailed soil survey of a 500 ha farmland resulted in the delineation of three main soil mapping units with the following general land characteristics:

Land Characteristics	Soil Mapping Units		
	A	**B**	**C**
Soil			
Texture	Loam	Gravelly Sandy Loam	Clay Loam
Bulk Density (surface) $(g.cm^{-1})$	1.8	1.5	1.4
Effective soil depth	150cm	20 cm to iron pan	50 cm to water table
pH (H_2O)	5.8	4.3	8.4
CEC (NH_4OAc)	4.9	5.5	9.2
Exch. Na (meq/100g)	0.08	0.10	2.10
Available P (ppm)	3.75	0.42	7.70
Electrical Conductivity (dSm^{-1})	0.90	0.04	4.25
Topography			
Slope (%)	2	8	0
Tree removal Required	Moderate	Moderate	Low
Drainage			
Internal Drainage	Well-drained	Excessively Well-drained	Seasonally water logged
Air Drainage	Good	Very good	Restricted

Use the information to answer the following questions:

(a)   What is a soil mapping unit?

(b) Discuss the general characteristics of each soil mapping unit separately.

(c) Compare the suitability of the three soil mapping units for the production of maize and swamp rice.

(d) Highlight the major problems associated with each of the soils and suggest possible solutions aimed at improving the arability of the soils.

References

Esu, I. E. 1982. *Evaluation of soils for irrigation in the Kaduna area of Nigeria.* Unpubl. Ph.D thesis. Dept. of Soil Sciecne, Ahmadu Bello University, Zaria. 305pp.

Esu, I. E. 1988. *Detailed soil survey of the Tsiga Scheme, Jare River Basin Irrigation project* (Final Report). A consultancy project commissioned by Enplan Group Consulting Engineers and Planners for the Katsina State Government of Nigeria. 86pp.

Dent, D. and A. Young. 1981. *Soil Survey and land evaluation.* Allen and Unwin, London.

FAO, 1976. *A Framework for Land Evaluation.* Soils Bull. No. 32, FAO, Rome, 79pp.

FAO, 1979. *Soil Survey Investigations for Irrigation.* Soils Bull. No. 42. FAO, Rome. 188pp.

FAO, 1984. Guidelines: Land Evaluation for Rainfed Agriculture. *Soils Bull.* No. 52, FAO, Rome. 237pp.

Klingebiel, A. A. and P. H. Montgomery. 1966. Land capability classification. USDA Soil Conservation Service. Agric. *Handbook No. 210, 21pp.*

Olson, G. W. 1974. Land classification. *Search Agriculture,* 4(7): 1-34.

Requier, J. 1971. *The parametric method of land evaluation.* FAO, Rome.

Storie, R. E. 1937. *An index for rating the agricultural value of soils.* Univ. Calif. Agric. Expt. Stn., Berkeley, California, USA.

United States Bureau of Reclamation. 1953. *Land Classification Handbook*. United States Dept. of Interior. Publ. V. Part 2. 53pp.

Verheye, W. M. 1986. Principles of land appraisal and land use planning within the European Community. *Soil Use Management* 2: 120-124.

Appendix 1: Pictorial Representation of Modal Profile of the Twelve Soil Orders of Soil Taxonomy (Source: USDANRCS,1999)

Entisols are soils that show little or no evidence of pedogenic horizon development.

Entisols occur in areas of recently deposited parent materials or in areas where erosion or deposition rates are faster than the rate of soil development; such as dunes, steep slopes, and flood plains. They occur in many environments.

ENTISOLS MAKE UP ABOUT 16% OF THE WORLD'S ICE-FREE LAND SURFACE.

Inceptisols are soils of semiarid to humid environments that generally exhibit only moderate degrees of soil weathering and development.

Inceptisols have a wide range in characteristics and occur in a wide variety of climates.

Gelisols are soils that have permafrost near the soil surface and/or have evidence of cryoturbation (frost churning) and/or ice segregation.

Gelisols are common in the higher latitudes or at high elevations.

Andisols form from weathering processes that generate mineral with little orderly crystalline structure. These minerals can result in an unusually high water- and nutrient-holding capacity.

As a group, Andisols tend to be highly productive soils. They include weakly weathered soils with much volcanic glass as well as more strongly weathered soils. They are common in cool areas with moderate to high precipitation, especially those areas associated with volcanic materials.

Histosols have a high content of organic matter and no permafrost. Most are saturated year round, but a few are freely drained. Histosols are commonly called bogs, moors, peats, or mucks.

Histosols form in decomposed plant remains that accumulate in water, forest litter, or moss faster than they decay. If these soils are drained and exposed to air, microbial decomposition is accelerated and the soils may subside dramatically.

Aridisols are soils that are too dry for the growth of mesophytic plants. The lack of moisture greatly restricts the intensity of weathering processes and limits most soil development processes to the upper part of the soils. Aridisols often accumulate gypsum, salt, calcium carbonate, and other materials that are easily leached from soils in more humid environments.

Aridisols are common in the deserts of the world.

Vertisols have a high content of expanding clay minerals. They undergo pronounced changes in volume with changes in moisture. They have cracks that open and close periodically, and that show evidence of soil movement in the profile.

Because they swell when wet, vertisols transmit water very slowly and have undergone little leaching. They tend to be fairly high in natural fertility.

Mollisols are soils that have a dark colored surface horizon relatively high in content of organic matter. The soils are base rich throughout and therefore are quite fertile.

Mollisols characteristically form under grass in climates that have a moderate to pronounced seasonal moisture deficit. They are extensive soils on the steppes of Europe, Asia, North America, and South America.

Alfisols are in semiarid to moist areas.

These soils result from weathering processes that leach clay minerals and other constituents out of the surface layer and into the subsoil, where they can hold and supply moisture and nutrients to plants. They formed primarily under forest or mixed vegetative cover and are productive for most crops.

Ultisols are soils in humid areas. They formed from fairly intense weathering and leaching processes that result in a clay-enriched subsoil dominated by minerals, such as quartz, kaolinite, and iron oxides.

Ultisols are typically acid soils in which most nutrients are concentrated in the upper few inches. They have a moderately low capacity to retain additions of lime and fertilizer.

Spodosols formed from weathering processes that strip organic matter combined with aluminum (with or without iron) from the surface layer and deposit them in the subsoil. In undisturbed areas, a gray eluvial horizon that has the color of uncoated quartz overlies a reddish brown or black subsoil.

Spodosols commonly occur in areas of coarse-textured deposits under coniferous forests of humid regions. They tend to be acid and infertile.

Oxisols are highly weathered soils of tropical and subtropical regions. They are dominated by low activity minerals, such as quartz, kaolinite, and iron oxides. They tend to have indistinct horizons.

Oxisols characteristically occur on land surfaces that have been stable for a long time. They have low natural fertility as well as a low capacity to retain additions of lime and fertilizer.

Appendix 2: Glossary of Pedological Terms

ABC soil. A soil having an A, a B, and a C horizon.

AC soil. A soil having only an A and a C horizon. Commonly, such soil formed in recent alluvium or on steep, rocky slopes.

Aeration, soil. The exchange of air in soil with air from the atmosphere. The air in a well aerated soil is similar to that in the atmosphere; the air in a poorly aerated soil is considerably higher in carbon dioxide and lower in oxygen.

Aggregate, soil. Many fine particles held in a single mass or cluster. Natural soil aggregates, such as granules, blocks, or prisms, are called peds. Clods are aggregates produced by tillage or logging.

Alkali (sodic) soil. A soil having so high a degree of alkalinity (pH 8.5 or higher) or so high a percentage of exchangeable sodium (15 percent or more of the total exchangeable bases), or both, that plant growth is restricted.

Alluvial fan. The fanlike deposit of a stream where it issues from a gorge upon a plain or a tributary stream near or at its junction with its main stream.

Alluvial plain. A flood plain or a low-gradient delta. It may be modern or relict.

Alluvium. Material, such as sand, silt, or clay, deposited on land by streams.

Aquic conditions. Current soil wetness characterized by saturation, reduction, and redoximorphic features.

Argillic or Argic horizon. A subsoil horizon characterized by an accumulation of illuvial clay.

Aspect. The direction in which a slope faces.

Association, soil. A group of soils or miscellaneous areas geographically associated in a characteristic repeating pattern and defined and delineated as a single map unit.

Available water capacity (available moisture capacity). The capacity of soils to hold water available for use by most plants. It is commonly defined as the difference between the amount of soil water at field moisture capacity and the amount at wilting point. It is commonly expressed as inches of water per inch of soil. The capacity, in inches, in a 60-inch profile or to a limiting layer is expressed as: Very low 0 to 2.5.

Low 2.5 to 5
Moderate 5 to 7.5
High 7.5 to 10
Very high more than 10

Back slope. The geomorphic component that forms the steepest inclined surface and principal element of many hillsides. Back slopes in profile are commonly steep, are linear, and may or may not include cliff segments.

Base saturation. The degree to which material having cation-exchange properties is saturated with exchangeable bases (sum of Ca, Mg, Na, K), expressed as a percentage of the total cation-exchange capacity.

Bedrock. The solid rock that underlies the soil and other unconsolidated material or that is exposed at the surface.

Bisequum. Two sequences of soil horizons, each of which consists of an illuvial horizon and the overlying eluvial horizons.s

Boulders. Rock fragments larger than 2 feet (60 centimeters) in diameter.

Calcareous soil. A soil containing enough calcium carbonate (commonly combined with magnesium carbonate) to effervesce visibly when treated with cold, dilute hydrochloric acid.

Capillary water. Water held as a film around soil particles and in tiny spaces between particles. Surface tension is the adhesive force that holds capillary water in the soil.

Cation. An ion carrying a positive charge of electricity. The common soil cations are calcium, potassium, magnesium, sodium,a nd hydrogen.

Cation-exchange capacity (CEC). The total amount of exchangeable cations that can be held by the soil, expressed in terms of milliequivalents per 100 grams of soil at neutrality (pH 7.0) or at some other stated pH value. The term, as applied to soils, is synonymous with base-exchange capacity but is more precise in meaning.

Clay. As a soil separate, the mineral soil particles less than 0.002 millimeter or 2 micrometer (2μ) in diameter. As a soil textural class, soil material that is 40 percent or more clay, less than 45 percent sand, and less than 40 percent silt.

Clay depletions. Low-chroma zones having a low content of iron, manganese, and clay because of the chemical reduction of iron and manganese and the removal of iron, manganese, and clay. A type of redoximorphic depletion.

Clayey soil. Silty clay, sandy clay, or clay.

Clay film. A thin coating of oriented clay on the surface of a soil aggregate or lining pores or root channels. Synonyms: clay coating, clay skin.

Claypan. A slowly permeable soil horizon that contains much more clay than the horizons above it. A claypan is commonly hard when dry and plastic or stiff when wet.

Coarse fragments. Mineral or rock particles larger than 2 millimeters in diameter.

Coarse textured soil. Sand or loamy sand.

Cobble (or cobblestone). A rounded ro partly rounded fragment of rock 13 to 10 inches (7.6 to 25 centimeters) in diameter.

Cobbly soil material. Material that is 15 to 35 percent, by volume, rounded or partially rounded rock fragments 3 to 10 inches (7.6 to 25 centimeters) in diameter. Very cobbly soil material is 35 to 60 percent of these rock fragments, and extremely cobbly soil material is more than 60 percent.

COLE (Coefficient of linear extensibility). See Linear Extensibility.

Colluvium. Soil material or rock fragments, or both, moved by creep, slide, or local wash and deposited at the base fo steep slopes.

Complex slope. Irregular or variable slope. Planning or establishing terraces, diversions, and other water-control structures on a complext slope is difficult.

Complex, soil. A map unit of two or more kinds of soil or miscellaneous areas in such an intricate pattern or so small in area that it is not practical to map them separately at the selected scale of mapping. The pattern and proportion of tehs oils or miscellaneous areas are somewhat similar in all areas.

Concretions. Cemented bodies with crude internal symmetry organized around a point, a line, or a plane that typically takes the form of concentric layers visible to the naked eye. Calcium carbonate, iron oxide, and manganese oxide are common compounds making up concretions. If formed in place, concretions of iron oxide or manganese oxide are generally considered a type of redoximorphic concentration.

Conglomerate. A coarse grained, classtic rock composed of rounded to suangular rock fragments more than 2 millimeters in diameter. It commonly has a matrix of sand and finer textured material. Conglomerate is the consolidated equivalent of gravel.

Consistence, soil. Refers to the degree of cohesion and adhesion of soil material and its resistance to deformation when ruptured. Consistence includes resistance of soil material to rupture and to penetration;l plasticity, toughness,a dn stickiness fo puddle soil material; and the manner in which the soil material behaves when subject to compression. Terms describing consistence are defined in this book.

Consolidated sandstone. Sandstone that disperse within a few hours when fragments are placed in water. The fragments are extremely hard or very hard when dry, are not easily crushed, and cannot be textured by the usual field method.

Consolidated shale. Shale that disperses within a few hours when fragments are placed in water. The fragments are extremely hard or very hard when dry and are not easily crushed.

Control section. The part of the soil on which classification is based. The thickness varies among different kinds of soil, but for many it is that part of the soil profile between depths of 25 cm and 100 cm or 200 cm.

Cuesta. A hill or ridge that has a gentle slope on one side and a steep slope on the other; specifically, an asymmetric, homoclinal ridge capped by resistant rock layers of slight or moderate dip.

Deep soil. A soil that is 40 to 60 inches or 100 to 150 cm deep over bedrock or to other material that restricts the penetration of plant roots.

Dense layer (in tables). A very firm, massive layer that has a bulk density of more than 1.8 grams per cubic centimeter. Such a layer affects the ease of digging and can affect filling and compacting.

Depth, soil. Generally, the thickness of the soil over bedrock. Very deep soils are more than 60 inches (150cm) deep over bedrock; deep soils, 40 to 60 inches (100-150cm); moderately deep, 20 to 40 inches (50-100cm); shallow, 10 – 20 inches (25-50cm); and very shallow, less than 10 inches (25cm).

Depth to rock (in tables). Bedrock is too near the surface for the specified use.

Drainage class (natural). Refers to the frequency and duration of wet periods under conditions similar to those under which the soil formed. Alterations of the water regime by human activities, either through drainage or irrigation, are not a consideration unless they have significantly changed the morphology of the soil. Seven classes of natural soil drainage are recognized – *excessively drained, somewhat excessively drained, well drained, moderately well drained, somewhat poorly drained, poorly drained, and very poorly drained.* These classes are defined in the "Soil Survey Manual".

Drainage, surface. Runoff, or surface flow of water, from an area.

Eluviation. The movement of material in true solution or colloidal suspension from one place to another within the soil. Soil horizons that have lost material through eluviation are eluvial; those that have received material are illuvial.

Endosaturation. A type of saturation of the soil in which all horizons between the upper boundary of saturation and a depth of 2 meters are saturated.

Episaturation. A type of saturation indicating a perched water table in a soil in which saturated layers are underlain by one or more unsaturated layers within 2 meters of the surface.

Erosion. The wearning away of the land surface by water, wind, ice, or other geologic agents and by such processes as gravitational creep.

Erosion (geologic). Erosion caused by geological processes acting over long geologic periods and resulting in the wearing away of mountains and the building up of such landscape features as flood plains and coastal plains. Synonym: natural erosion.

Erosion (accelerated). Erosion much more rapid than geologic erosion, mainly as a result of human or animal activities or of a catastrophe in nature, such as a fire, that exposes the surface.

Erosion pavement. A layer of gravel or stoens that remains on the surface after fine particles are removed by sheet or rill erosion.

Extrusive rock. Igneous rock derived from deep-seated molten matter (magma) emplaced on the earth's surface.

Fertility, soil. The quality that enables a soil to provide plant nutrients, in adequate amounts and in proper balance, for the growth of specified plants when light, moisture, temperature, tilth, and other growth factors are favorable.

Field moisture capacity. The moisture content of a soil, expressed as a percentage of the ovendry weight, after the gravitational, or free, water has drained away; the field moisture content 2 or 3 days after a soaking rain; also called *normal field capacity, normal moisture capacity,* or *capillary capacity.*

Fine textured soil. Sandy clay, silty clay, or clay.

Flood plain. A nearly level alluvial plain that borders a stream and is subject to flooding unless protected artificially.

Fluvial. Of or pertaining to rivers; produced by river action, as a fluvial plain.

Foot slope. The inclined surface at the base of a hill.

Fragile (in tables). A soil that is easily damaged by use or disturbance.

Genesis, soil. The mode of origin of the soil. Refers especially to the processes or soil-forming factors responsible for the formation of the solum, or true soil, from the unconsolidated parent material.

Gilgai. Commonly, a succession of microbasins and microknolls in nearly level areas or of microvalleys and microridges parallel with the slope. Typically, the microrelief of clayey soils that shrink and swell considerably with changes in moisture content.

Gleyed soil. Soil that formed under poor drainage, resulting in the reduction of iron and other elements in the profile and in gray colors.

Gravel. Rounded or angular fragments of rock as much as 3 inchs (2 millimeters to 7.6 centimeters) in diameter. An individual piece is a pebble.s

Gravelly soil material. Material that is 15 to 50 percent, by volume, rounded or angular rock fragments, not prominently flattened, as much as 3 inches (7.6 centimeters) in diameter.

Ground water. Water filling all the ublocked pores of underlying material below the water table.

Gully. A miniature valley with steep sides cut by running water and through which water ordinarily runs only after rainfall. The distinction between a gully and a rill is one of depth. A gully generally is an obstacle to farm machinery and is too deep to be obliterated by ordinary tillage; a rill is of lesser depth and can be smoothed over by ordinary tillage.

Gypsum. A mineral consisting of hydrous calcium sulfate.

Hard bedrock. Bedrock that cannot be excavated except by blasting or by the use of special equipment that is not commony used in construction.

Hardpan. A hardened or cemented soil horizon, or layer. The soil material is sandy, loamy, or clayey and is cemented by iron oxide, silica, calcium carbonate, or other substance.

Hill slope. The steeper part of a hill between its summit and the drainage line, valley flat, or depression floor at the base of the hill. In descending order, geomorphic components of a simple hill slope may include shoulder, back slope, foot slope,a nd toe slope. However, all of these components are not necessarily present in any given hill slope continuum. In addition, complex hill slopes may include two or more back slope to toe slope sequences.

Horizon, soil. A layer of soil, approximately parallel to the surface, having distinct characateristics produced by soil-forming processes. In the identification of soil horizons, an uppercase letter represents the major horizons. Numbers or lowercase letters that follow represent subdivisions of the major horizons. The major horizons of mineral soil are as follows:

O horizon.- An organic layer of fresh and decaying plant residue.

A horizon. – The mineral horizon at or near the surface in which an accumulation of humified organic matter is mixed with the mineral material. Also, a plowed surface horizon, most of which was originally part of a B horizon.

E horizon. – The mineral horizon in which the main feature is loss of silicate clay, iron, aluminum, or some combination of these.

B horizon. – The mineral horizon below an A horizon. The B horizon is in part a layer of transition from the overlying A to the underlying C horizon. The B horizon also has distinctive characateristics, such as (1) accumulation of clay, sessquixides, humus, or a combination fo these; (2) prismatic or bocky structure; (3) redder or browner colors than those in the A horizon; or (4) a combination of these.

C horizon.- The mineral horizon or layer, excluding indurated bedrock, that is little affected by soil-forming processes and does not have the properties typical of the overlying soil material. The material of a C horizon may be either like or unlike that in which the solum formed. If the material is known to differ from that in the soum, an Arabic numeral, commonly a 2, precedes the letter C.

Cr horizon. – Soft, consolidated bedrock beneath the soil.

R layer. – Consolidated bedrock beneath the soil. The bedrock commonly nderlies a C horizon, but it can be directly below an A or a B horizon.

Hummock. A rounded or conical mound or knoll, hillock, ro other small elevation. Also, a slight rise of ground above a level surface.

Humus. The well decomposed, more or less stable part of the organic matter in mineral soils.s

Igneous rock. Rock formed by solidification from a molten or partially molten state. Major varieties include plutonic and volcanic rock. Examples are andesite, baslt, and granite.

Illuviation. The movement of soil material from one horizon to another in the soil profile. Generally, material is removed from an upper horizon and deposited in a lower horizon.

Infiltration. The downward entry of wate into the immediate surface of soil ro other material, as contrasted with percolation, which is movement of water through soil layers or material.

Infiltration capacity. The maximum rate at which water can infiltrate into a soil under a given set of conditions.

Infiltration rate: The average rate of water entering the soil under irrigation. Most soils have a fast initial rate; the rate decreases with application time. Therefore, intake rate for design purposes is not a constant but is a variable depending on the net irrigation application. The rate of water intake, in inches per hour, is expressed as follows:
    Less than 0.2 very low
    0.2 to 0.4 low
    0.4 to 0.75 moderately low
    0.75 to 1.25 moderate
    1.25 to 1.75 moderately high

1.75 to 2.5 high
More than 2.5 very high

Irrigation. Application of wate to soils to assist in production of crops. Methods of irrigation are:

Basin. – Water is applied rapidly to nearly level plains surrounded by levees or dikes.

Border. – Water is applied at the upper end fo a strip in which the lateral flow of water is controlled by samlle earth ridges called border dikes, or borders.

Controlled flooding. - Water is released at intervals from closely spaced field ditches and distributed uniformly over the field.

Corrugation. – Water is applied to small, closely spaced furrows or ditches in fields of close-growing crops or in orchards so that it flows in only one direction.

Drip (or trickle). – Water is applied slowly and under low pressure to the surface of the soil or into the soil through such applicators as emitters, prous tubing, or perforated pipe.

Furrow. – Water is applied in small ditches made by cultivation implements. Furrows are used for tree and row crops.

Sprinkler: - Water is sprayed over the soil surface through pipes or nozzles from a pressure system.

Subirrigation. – Water is applied in open ditches or tile lines until the water table is raised enough to wet the soil.

Wild flooding. – Water, released at high points, is allowed to flow onto an area without controlled distribution.

Knoll. A small, low, rounded hill rising above adjacent landforms.
Lacustrine deposit. Material deposited in lake water and exposed when the water level is lowered or the elevation of the land is raised.

Leaching. The removal of soluble material from soil or other material by percolating water.

Linear Extensibility Percent (LEP). The linear expression of the volume difference of natural soil fabric at 1/3 bar or 1/10 bar water content and oven dryness. The volume change is reported as percent change for the whole soil.

Liquid limit. The moisture content at which the soil passes from a plastic to a liquid state.

Loam. Soil material that is 7 to 27 percent clay particles, 28 to 50 percent silt particles, and less than 52 percent sand particles.

Loamy soil. Coarse sandy loam, sandy loam, fine sandy loam, very find sandy loam, loam, silt loam, silt, clay loam, sandy clay loam, or silty clay loam.

Medium textured soil. Very fine sandy loam, loam, silt loam, or silt.

Metamorphic rock. Rock of any origin altered in mineralogical composition, chemical composition, or structure by heat, pressure, and movement. Nearly all such rocks are crystalline.

Metamorphism. The mineralogical and structural adjustment of solid rocks to physical and chemical conditions which have been imposed at depth below the surface zones of weathering and cementation.

Mineral soil. Soil that is mainly mineral material and low in organic material. Its bulk density is more htan that of organic soil.

Miscellaneous area. An area that has little or no natural soil and supports little or no vegetation.

Mollic epipedon. A thick, dark, humus-rich surface horizon (or horizons) that has high base saturation and pedogenic soil structure. It may include the upper part of the subsoil.

Morphology, soil. The physical makeup of the soil, including the texture, structure, porosity, consistence, color, and other physical, mineral, and biological properties of the various horizons, and the thickness and arrangement of those horizons in the soil profile.

Mottling, soil. Irregular spots of different colors that vary in number and size. Descriptive terms are as follows: abundance – *few, common,* and *many; size – fine, medium,* and *coarse;l* and contrast – *faint, distinct,* and *prominent.* The size measurements are of the diameter along the greatest dimension. *Fine* indicates less than 5 millimeters (about 0.2 inch); *medium,* from 5 ot 15 millimeters (about 0.2 to 0.6 inch); and *coarse,* more than 15 millimeters (about 0.6 inch).

Mountain. A natural elevation fo the land surface, rising more htan 1,000 feet above surrounding lowlands, commonly of restricted summit area (relative to a plateau) and generally having steep sides. A mountain can occur as a single, isolated mass or in a group forming a chain or range.

Mudstone. Sedimentary rock formed by induration fo silt and clay in approximately equal amounts.

Munsell notation. A designation of color by degrees of three simple variables – hue, value,a nd chroma. For example, a notation of 10YR 6/4 is a color with hue of 10YR, value of 6, and chroma of 4.

Natric horizon. A special kind of argillic horizon that contains enough exchangeable sodium to have an adverse effect on the physical condition of the subsoil.

Neutral soil. A soil having a pH value between 6.6 and 7.3 (See Reaction, soil.)

Nodules. Cemented bodies lacking visible internal structure. Calcium carbonate, iron oxide, and manganese oxide are common compounds making up nodules. If formed in place, nodules of iron oxide or manganese oxide are considered types of redoximorphic concentrations.

Nutrient, plant. Any element taken in by a plant essential to its growth. Plant nutrients are mainly nitrogen, phosphorus, potassium, calcium, magnesium, sulfur, iron, manganese, copper, boron, and zinc obtained from the soil and carbon, hydrogen, and oxygen obtained from the air and water.

Organic matter. Plant and animal residue in the soil in various stages of decomposition. The content of organic matter in the surface layer is described as follows:
Very low less than 0.5 percent
Low 0.5 to 1.0 percent
Moderately low 1.0 to 2.0 percent
Moderate 2.0 to 4.0 percent
high 4.0 to 8.0 percent
Very high more than 8.0 percent

Pan. A compact, dense layer in a soil that impedes the movement of water and the growth of roots. For example, *hardpan, fragipan, claypan, plowpan,* and *traffic pan.*

Parent material. The unconsolidated organic and mineral material in which soil forms.

Ped. An individual natural soil aggregate, such as a granule, a prism, or a block.

Pediment. A gently sloping erosional surface developed at the foot of a receding hill or mountain slope. The surface may be essentially bare, exposing earth material that extends beneath adjacent uplands; or it may be thinly mantled with alluvium and colluvium.

Pedon. The smalles t volume that can be called "a soil". A pedon is three dimensional and large enough to permit study of all horizons. Its area ranges from about 10 to 100 square feet (1 square meter to 10 square meters), depending on the variability of the soil.

Phase, soil. A subdivision on a soil series based on features that affect its use and management, such as slope, stoniness, and flooding.

pH value. A numerical designation fo acidity and alkalinity in soil. (See Reaction, soil).

Plasticity index. The numerical difference between the liquid limit and the plastic limit; the range of moisture content within which the soil remains plastic.

Plasitc limit. The moisture content at which a soil changes from semisolid to plastic.

Plowpan. A compacted layer formed in the soil directly below the plowed layer.

Ponding. Standing water on soils in closed depressions. Unless the soils are artificially drained, the water can be removed only by percolation or evapotranspiration.

Potential rooting depth (effective rooting depth). Depth to which roots could penetrate if the content of moisture in the soil were adequate. Th soil has no properties restricting the penetration of roots to this depth.

Productivity, soil. The capability of a soil for producing a specified plant or sequence of planats under specific management.

Profile, soil. A vertical section of the soil extending through all its horizons and into the parent material.

Quartzite, metamorphic. Rock consisting mainly of quartz that formed through recrystallization of quartz-rich sandstone or chert.

Reaction, soil. A measure of acidity or alkalinity of a soil, expressed in pH values. A soil that tests to pH 7.0 is described as precisely neutral in reaction because it is neither acid nor alkaline. The degrees of acidity or alkalinity, expressed as pH values, are:

Ultra acid less than 3.5
Extreme acid 3.5 to 4.4
Very srongly acid 4.5 to 5.0
Strongly acid 5.1 to 5.5
Moderately acid 5.6 to 6.0
Slightly acid 6.1 to 6.5
Neutral 6.6 to 7.3
Slightly alkaline 7.4 to 7.8
Moderately alkaline 7.9 to 8.4
Strongly alkaline 8.5 to 9.0
Very strongly alkaline 9.1 and higher

Redoximorphic concentrations. Nodules, concretions, soft masses, pore linings, and other features resulting from the accumulation of iron or manganese oxide. An indication of chemical reduction and oxidation resulting from saturation.

Redoximorphic depletions. Low-chroma zones from which iron and manganese oxide or a combination of iron and manganese oxide and clay has been removed. These zones are indications of the chemical reduction of iron resulting from saturation.

Redoximorphic features. Redoximorphic concentrations, redoximorphic depletions, reduced matrices, a positive reaction to alpha, alpha-dipyridyl, and other features indicating the chemical reduction and oxidation of iron and manganese compounds resulting from saturation.

Reduced matrix. A soil matrix that has low chroma in situ because of chemically reduced iron (Fe II). The chemical reduction results from nearly continuous wetness, the matrix undergoes a change in hue or chroma within 30 minutes after exposure to air as the iron is oxidized (Fe III). A type of redoximorphic feature.

Regolith. The unconsolidated mantle of weather rock and soil material on the earth's surface; the loose earth material above the solid rock.

Relief. The elevations or inequalities of a land surface, considered collectively.

Residuum (residual soil material). Unconsolidated, weathered or partly weathered mineral material that accumulated as consolidated rock disintegrated in place.

Ridge. A long, narrow elevation of the land surface, usually sharp crested with steep sides and forming an extended upland between valleys. The term is used in areas of both hill and mountain relief.

Rill. A steep-sided channel resulting from accelerated erosion. A rill is generally a few inches deep and not wide enough to be an obstacle to farm machinery.

Road cut. A sloping surface produced by mechanical means during road construction. It is commonly on the uphill side of the road.

Rock fragments. Rock or mineral fragments having a diameter of 2 millimeters or more; for example, pebbles, cobbles, stones, and boulders.

Rock outcrop. Exposures of bare bedrock other than lava flows and rock-lined pits.

Runoff. The precipitation discharged into stream channels from an area. The water that flows off the surface of the land without sinking into the soil is called surface runoff. Water that enters the soil before reaching surface streams is called ground-water runoff or seepage flow from ground water.

Saline soil. A soil containing soluble salts in an amount that impairs the growth of plants. A saline soil does not contain excess exchangeable sodium.

Salinity. The electrical conductivity of a saline soil. It si expressed, decisimens permeter (dSm^{-1}), as follows

Nonsaline 0 to 2
Slighty saline 2 to 4
Moderately saline 4 to 8
Strongly saline More than 8

Sand. As a soil separate, individual rock or mineral fragments from 0.05 millimeter to 2.0 millimeters in diameter. Most sand grains consist of quartz. As a soil texturela class, a soil that is 85 percent or more sand and not more than 10 percent clay.

Sandstone. Sedimentary rock containing dominantly sand-sized particles.

Sandy soil. Sand or loamy sand.

Saturation. Wetness characterized by zero or positive pressure of the soil water. Under conditions of saturation, the water will flow from the soil matrix into an unlined auger hole.

Sedimentary plain. An extensive nearly level to gently rolling or moderately sloping area that is underlain by sedimentary bedrock and that has a slope of 0 to 8 percent.

Sedimentary rock. Rock made up of particles deposited from suspension in water. The chief kinds of sedimentary rock are conglomerate, formed from gravel; sandstone, formed from sand; shale, formed from clay; and limestone, formed from soft masses of calcium carbonate. There are many intermediate types. Some wind-deposited sand is consolidated into sandstone.

Sequum. A sequence consisting of an illuvial horizon and the overlying eluvial horizon. (See Eluvation.)

Series, soil. A group of soils that have profiles that are almost alike, except for differences in texture of the surface layer. All the soils of a series have horizons that are similar in composition, thickness, and arrangement.

Shale. Sedimentary rock formed by the hardening of a clay deposit.

Shallow soil. A soil that is 10 to 20 inches deep over bedrock or to other material that restricts the penetration of plant roots.

Sheet erosion. The removal of a fairly uniform layer of soil material from the land surface by teha ction of rainfall and surface runoff.

Shoulder slope. The uppermost inclined surface at the top of a hillside. It si the transition zone from the back slope to the summit of a hill or mountain. The surface is dominantly convex in profile and erosional in origin.

Shrink-swell. The shrinking of soil when dry and the swelling when set. Shrinking and swelling can damange roads, dams, building foundations, and other structures. It can also damange plant roots.

Silica. A combination of silica and oxygen. The mineral form is called quartz.

Silica-sesquioxide ratio. The ratio of the number of molecules fo silica to the number of molecules fo alumina and iron oxide. The more highly weathered soils or their clay fractions in warm-temperate, humid regions,a dn especially those in the tropics, generally have a low ratio.

Silt. As a soil separate, individual mineral particles that range in diameter form the upper limit of clay (0.002 millimeter) to the lower limit of very fine sand (0.05 millimeter). As a soil textural class, soil that is 80 percent or more silt and less than 12 percent clay.

Siltstone. Sedimentary rock made up of dominantly silt-sized particles.

Slickensides. Polished and grooved surfaces produced by one mass sliding past another. In soils, slickensides may occur at the bases of slip surfaces on the steeper slopes; on faces of blocks, prisms, and columnsl; and in swelling clayey soils, where there is marked change in moisture content.

Slope. The inclination of the land surface from the horizontal. Percentage of slope is the vertical distance divided by horizontal distance, then multiplied by 100. Thus, a slope of 20 percent is a drop of 20 feet in 100 fet of horizontal distance. The following slope classes are often recognized.

Nearly level 0 to 2 percent
Gently sloping 2 to 5 percent
Moderately sloping 5 ot 9 percent
Strongly sloping 9 to 15 percent
Moderately steep 15 to 30 percent
Steep 30 to 50 percent
Very steep 50 percent and higher

Classes for complex slopes are as follows:

Nearly level 0 to 2 percent
Undulating 2 to 5 percent
Gently rolling 5 to 9 percent
Rolling 9 to 15 percent
Hilly 15 to 30 percent
Steep 30 to 50 percent
Very steep 50 percent and higher

Sodic (alkali) soil. A soil having so high a degree of alkalinity (pH 8.5 or higher) or so high a percentage of exchangeable sodium (15 percent or more of the total exchangeable bases), or both, that plant growth is restricted.

Sodicity. The degree to which a soil is affected by exchangeable sodium. Sodicity is expressed as a sodium adsorption ratio (SAR) of a saturation extract, or the ratio of sodium to calcium plus magnesium. The degrees of sodicity and their respective ratios are:

Slight less than 13:1
Moderate 13-30:1
Strong more than 30:1

Soft bedrock. Bedrock that can be excavated with trenching machines, backhoes, small rippers, and othe equipment commonly used in construction.

Soil. A natural, three-dimensional body at the earth's surface. It is capable of supporting plants and has properties resulting from the integrated effect of climate and living matter acting on earthly parent material, as conditioned by relief over periods of time.

Soil separates. Mineral particles less than 2 millimeters in equivalent diameter and ranging between specified size limits. The names and sizes, in millimeters, of separates recognized in the United States are as follows:

Very coarse sand 2.0 to 1.0
Coarse sand 1.0 to 0.5
Medium sand 0.5 to 0.25
Fine sand 0.25 to 0.10
Very fine sand 0.10 to 0.05
Silt 0.05 to 0.002
Clay less than 0.002

Solum. The upper part of a soil profile, above the C horizon, in which the processes of soil formation are active. The solum in soil consists fo the A, E, and B horizons. Generally, the characteristics of the material in these horizons are unlike those of the material below the solum. The living roots and plant and animal activities are largely confined to the solum.

Stones. Rock fragments 10 to 24 inches (25 to 60 centimeters) in diameter if rounded or 15 ot 24 inches (38 to 60 centimeters) in length if flat.

Stony. Refers to a soil containing stones in numbers that interfere with or prevent tillage.

Structure, soil. The arrangement of primary soil particles into compound particles or aggregates. The principal forms of soil structure are: *platy* (laminated), *prismatic* (vertical axis of aggregates longer than horizontal), *columnar* (prisms with rounded tops), *blocky* (angular or subangular), and *granular*. *Structureless* soils are either *single grain* (each grain by itself, as in dune sand) or *massive* (the particles adhering without any regular cleavage, as in many hardpans).

Subsoil. Technically, the B horizon; roughly, the part of the solum below plow depth.

Subsoiling. Tilling a soil below normal plow depth, ordinarily to shatter a hardpan or claypan.

Substratum. The part of the soil below the solum.

Summit. A general term for the top, or highest level, of an upland feature, such as a hill or mountain. It commonly refers to a higher aera that has a gentle slope and is flanked b steeper slopes.

Surface layer. The soil ordinarily moved in tillage, or its equivalent in uncultivated soil, ranging in depth 10 to 25 centimeters. Frequently designated as the "plow layer", or the "Ap horizon".

Surface soil. The A, E, AB, and EB horizons, considered collectively. It includes all subdivisions of these horizons.

Taxadjuncts. Soils that cannot be classified in a series recognized in the classification system. Such soils are named for a series they strongly resemble and are designated as taxadjuncts to that series because they differ in ways too small to be of consequence in interpreting their use and behaviour. Soils are recognized as taxadjuncts only when one or more of their characteristics are slightly outside the range defined for the family of the series for which the soils are named.

Terrace (geologic). An old alluvial plain, ordinarily flat or undulating, bordering a river, a lake, or the sea.

Texture, soil. The relative proportions of sand, silt, and clay particles in a mass of soil. The basic textural classes, in order of increasing proportion of fine particles, are *sand, loam sand, sandy loam, loam, silt loam, silt, sandy clay loam, clay loam, silty clay loam, sandy clay, silty clay,* and *clay*. The sand, loamy sand, and sandy loam classes may be further divided by specifying "coarse", "fine", or "very fine".

Toe slope. The outermost inclined surface at the base of a hill; part of a foot slope.

Topsoil. The upper part of the soil, which is the most favorable material for plant growth. It si ordinarily rich in organic matter and is used to topdress roadbanks, lawns, and land affected by mining.

Toxicity (in tables). Excessive amount of toxic substances, such as sodium or sulfur, that severely hinder establishment of vegetation or severely restrict plant growth.

Trace elements. Chemical elements, for example, zinc, cobalt, manganese, copper, and iron, in soils in extremely small amounts. They are essential to plant growth.

Water table, perched. The surface of a local zone of saturation held above the main body of groundwater by an impermeable layer or stratum, usually clay, and separated from the main body of groundwater by an unsaturated zone.

Weathering. All physical and chemical changes produced in rocks or other deposits at or near the earth's surface by atmospheric agents. These changes result in disintegration and decomposition of the material.

Index